Innovation, Technology, and Knowledge Management

More information about this series at http://www.springer.com/series/8124

Igor N. Dubina • Elias G. Carayannis
Editors

Creativity, Innovation, and Entrepreneurship Across Cultures

Theory and Practices

Springer

Editors
Igor N. Dubina
Altai State University
Barnaul, Russia

Novosibirsk National Research State University
Novosibirsk, Russia

Elias G. Carayannis
School of Business
George Washington University
Washington, District of Columbia, USA

ISSN 2197-5698 ISSN 2197-5701 (electronic)
Innovation, Technology, and Knowledge Management
ISBN 978-1-4939-3260-3 ISBN 978-1-4939-3261-0 (eBook)
DOI 10.1007/978-1-4939-3261-0

Library of Congress Control Number: 2015958886

Springer New York Heidelberg Dordrecht London

Printed on acid-free paper

Springer Science+Business Media LLC New York is part of Springer Science+Business Media (www.springer.com)

Series Foreword

The Springer book series *Innovation, Technology, and Knowledge Management* was launched in March 2008 as a forum and intellectual, scholarly "podium" for global/local, transdisciplinary, transsectoral, public–private, and leading/"bleeding"-edge ideas, theories, and perspectives on these topics.

This book series is accompanied by the Springer *Journal of the Knowledge Economy*, which was launched in 2009 with the same editorial leadership.

This series showcases provocative views that diverge from the current "conventional wisdom," that are properly grounded in theory and practice, and that consider the concepts of ***robust competitiveness***[1], ***sustainable entrepreneurship***[2], and ***democratic capitalism***[3], central to its philosophy and objectives. More specifically, the aim of this series is to highlight emerging research and practice at the dynamic intersection of these fields, where individuals, organizations, industries, regions, and nations are harnessing creativity and invention to achieve and sustain growth.

---

[1] We define *sustainable entrepreneurship* as the creation of viable, profitable, and scalable firms. Such firms engender the formation of self-replicating and mutually enhancing innovation networks and knowledge clusters (innovation ecosystems), leading toward robust competitiveness (Carayannis, E. G. (2009). *International Journal of Innovation and Regional Development, 1*(3), 235–254).

[2] We understand *robust competitiveness* to be a state of economic being and becoming that avails systematic and defensible "unfair advantages" to the entities that are part of the economy. Such competitiveness is built on mutually complementary and reinforcing low-, medium-, and high-technology and public and private sector entities (government agencies, private firms, universities, and nongovernmental organizations) (Carayannis, E. G. (2009). *International Journal of Innovation and Regional Development, 1*(3), 235–254).

[3] The concepts of *robust competitiveness* and *sustainable entrepreneurship* are pillars of a regime that we call "*democratic capitalism*" (as opposed to "popular or casino capitalism"), in which real opportunities for education and economic prosperity are available to all, especially—but not only—younger people. These are the direct derivative of a collection of top-down policies as well as bottom-up initiatives (including strong research and development policies and funding, but going beyond these to include the development of innovation networks and knowledge clusters across regions and sectors) (Carayannis, E. G. & Kaloudis, A. (Jan 2009). *Japan Economic Currents*, p. 6–10).

Books that are part of the series explore the impact of innovation at the "macro" (economies, markets), "meso" (industries, firms), and "micro" levels (teams, individuals), drawing from related disciplines such as finance, organizational psychology, research and development, science policy, information systems, and strategy, with the underlying theme that for innovation to be useful it must involve the sharing and application of knowledge.

Some of the key anchoring concepts of the series are outlined in the figure below and the definitions that follow (all definitions are from Carayannis E. G. & Campbell, D. F. J. (2009). *International Journal of Technology Management*, *46*, 3–4).

- The "Mode 3" Systems Approach for Knowledge Creation, Diffusion, and Use: "Mode 3" is a multilateral, multinodal, multimodal, and multilevel systems approach to the conceptualization, design, and management of real and virtual, "knowledge-stock" and "knowledge-flow," modalities that catalyze, accelerate, and support the creation, diffusion, sharing, absorption, and use of cospecialized knowledge assets. "Mode 3" is based on a system-theoretic perspective of socioeconomic, political, technological, and cultural trends and conditions that shape the coevolution of knowledge with the "knowledge-based and knowledge-driven global/local economy and society."
- Quadruple Helix: Quadruple helix, in this context, means to add to the triple helix of government, university, and industry a "fourth helix" that we identify as the "media-based and culture-based public." This fourth helix associates with "media," "creative industries," "culture," "values," "lifestyles," "art," and perhaps also the notion of the "creative class."
- Innovation Networks: Innovation networks are real and virtual infrastructures and infratechnologies that serve to nurture creativity, trigger invention, and catalyze innovation in a public and/or private domain context (for instance, government–university–industry public–private research and technology development coopetitive partnerships).
- Knowledge Clusters: Knowledge clusters are agglomerations of cospecialized, mutually complementary, and reinforcing knowledge assets in the form of "knowledge stocks" and "knowledge flows" that exhibit self-organizing, learning-driven, dynamically adaptive competences and trends in the context of an open systems perspective.
- Twenty-First Century Innovation Ecosystem: A twenty-first century innovation ecosystem is a multilevel, multimodal, multinodal, and multiagent system of systems. The constituent systems consist of innovation metanetworks (networks of innovation networks and knowledge clusters) and knowledge metaclusters (clusters of innovation networks and knowledge clusters) as building blocks and organized in a self-referential or chaotic fractal knowledge and innovation architecture (Carayannis 2001), which in turn constitute agglomerations of human, social, intellectual, and financial capital stocks and flows as well as cultural and technological artifacts and modalities, continually coevolving, cospecializing, and cooperating. These innovation networks and knowledge clusters also form,

reform, and dissolve within diverse institutional, political, technological, and socioeconomic domains, including government, university, industry, and non-governmental organizations and involving information and communication technologies, biotechnologies, advanced materials, nanotechnologies, and next-Generation energy technologies.

Who is this book series published for? The book series addresses a diversity of audiences in different settings:

1. *Academic communities*: Academic communities worldwide represent a core group of readers. This follows from the theoretical/conceptual interest of the book series to influence academic discourses in the fields of knowledge, also carried by the claim of a certain saturation of academia with the current concepts and the postulate of a window of opportunity for new or at least additional concepts. Thus, it represents a key challenge for the series to exercise a certain impact on discourses in academia. In principle, all academic communities that are interested in knowledge (knowledge and innovation) could be tackled by the book series. The interdisciplinary (transdisciplinary) nature of the book series underscores that the scope of the book series is not limited a priori to a specific basket of disciplines. From a radical viewpoint, one could create the hypothesis that there is no discipline where knowledge is of no importance.
2. *Decision makers—private/academic entrepreneurs and public (governmental, subgovernmental) actors*: Two different groups of decision makers are being addressed simultaneously: (a) private entrepreneurs (firms, commercial firms, academic firms) and academic entrepreneurs (universities), interested in optimizing knowledge management and in developing heterogeneously composed knowledge-based research networks, and (b) public (governmental, subgovernmental) actors that are interested in optimizing and further developing their policies and policy strategies that target knowledge and innovation. One purpose of public *knowledge and innovation policy* is to enhance the performance and competitiveness of advanced economies.
3. *Decision makers in general*: Decision makers are systematically being supplied with crucial information, for how to optimize knowledge-referring and knowledge-enhancing decision-making. The nature of this "crucial information" is conceptual as well as empirical (case-study-based). Empirical information highlights practical examples and points toward practical solutions (perhaps remedies); conceptual information offers the advantage of further-driving and further-carrying tools of understanding. Different groups of addressed decision makers could be decision makers in private firms and multinational corporations, responsible for the knowledge portfolio of companies; knowledge and knowledge management consultants; globalization experts, focusing on the internationalization of research and development, science and technology, and innovation; experts in university/business research networks; and political scientists, economists, and business professionals.

4. *Interested global readership*: Finally, the Springer book series addresses a whole global readership, composed of members who are generally interested in knowledge and innovation. The global readership could partially coincide with the communities as described above ("academic communities," "decision makers") but could also refer to other constituencies and groups.

Washington, DC Elias G. Carayannis

Preface

At present, cross-cultural research is one of the most dynamically and rapidly growing kind of multidisciplinary integrative study which has demonstrated its importance and effectiveness in many theoretical and practical fields. At the same time, creativity, innovation, and entrepreneurship are championed in the literature as the critical element that is vital not just for companies but also for the development of societies. A sizable body of research demonstrates that cultural differences may foster or inhibit creative, inventive, innovative, and entrepreneurial activities, and each culture has its own strengths and weaknesses regarding them. Better understanding of cultural diversity in these phenomena can help integrate cultural realities in order to use their advantages, on the one hand, and, on the other, to overcome paralyzing cultural limitations and compensate for the limits of each culture to better foster creativity, innovation, and entrepreneurship in the increasingly globalized social, economic, and business environments.

Recent studies in this field represent a comparatively new class of interdisciplinary research. This is a field where cultural, sociological, psychological, historical, economic, management, technology, and business studies closely intersect. Cross-cultural interest in creativity, innovation, and entrepreneurship has spread widely, resulting in various and often contradictory findings and papers representing both theoretical and empirical research on the cultural impact of these phenomena.

In this book, we review, analyze, systematize, and discuss various concepts, assumptions, speculations, theories, and empirical research which focus on the effect of cultures (mainly national cultures) on creativity, innovation, and entrepreneurship. Some aspects relating to these phenomena may be common among cultures and are universally applicable, while others are strongly determined by the particular cultural factors that make them difficult to transfer into other cultural contexts. Thus, we follow a combined approach of universalism and particularism (Garci´a-Cabrera and Garci´a-Soto 2008).

In this monograph, we deal with the cultural aspects of creativity, innovation, and entrepreneurship. But when we talk about "cultural aspects," we definitely face a challenge of culture understanding. As Raymond Williams, one of the pioneers of cultural sociology, notes, the term "culture" is among the most complicated words (see Inkster 2007). Actually, people belong to many different "human groups" at the same time (a nation, a gender, a generation, an organization, etc.). Each of these groups has a culture, and each culture influences behavior (Hofstede 1980, 1991). This monograph primarily focuses on only one of these cultures—national culture, i.e., a set of collective beliefs and values that distinguish people of one nationality from those of another (Hofstede 1991).

Although a significant number of studies have been published in this field recently, it is not surprising that no mature theory has yet emerged encompassing the various cultural dimensions and their effect on creativity, invention, innovation, and entrepreneurship. Whereas notable progress is being made by scholars, the literature remains underdeveloped in terms of systematization and integration of the results of cross-cultural research on these phenomena. Clearly, cultural factors do impact creativity and inventiveness and their manifestation in innovation and entrepreneurship, but when it comes to differences across cultures, the picture is far from clear.

In this book, we consider creativity, invention, innovation, and entrepreneurship as closely interrelated phenomena. Etymologically, creativity means "producing something from nothing," and, most generally, creativity manifests itself in producing ideas or some content (intellectual, spiritual, emotional, physical, etc.) (Dubina 2000, 2009). One of the practical aspects of creativity is invention. Invention is broadly understood as a creative idea transformed into an applicable form. So, inventiveness can be considered as the ability to transform a creative idea into an applicable form (a scheme, plan, instruction, etc.) in the technical, economic, business, political, social, and other spheres. Innovation, consequently, can be considered as a practically implemented invention (innovation is a new idea that adds a value). Entrepreneurship, which is widely regarded as an engine of economic growth and development, is understood as creating a new way of doing business. At the same time, creativity is conceptualized "as a process of perceiving new relationships and new challenges, coping with changing situations, and expressing one's unique perceptions and responses" (Raina 1999). A circle has been enclosed.

The aim of this monograph is to further develop the relationship between culture and the manifold phenomena of creativity, innovation, and entrepreneurship in order to promote further and better understanding of how, why, and when these phenomena are manifested themselves across different cultures.

This book was prepared by a strong international and eminently qualified team. The authors present perspectives, insights, and contrasts from China, France, Greece, Iran, Italy, Japan, Russia, Singapore, the UK, and the USA. We hope that this book will be of interest and use for scholars and practitioners who closely deal with the theoretical and practical aspects of cultural and cross-cultural studies of creativity, innovation, and entrepreneurship, as well as international business and management.

I sincerely and greatly appreciate the valuable efforts of my respectful colleagues who kindly took up my initiative to prepare a monograph on such an interesting topic and submitted their chapters. I also express my deepest gratitude to Prof. Dr. Elias Carayannis, the Editor of the Springer Innovation, Technology, and Knowledge

Management Book Series and Coeditor of this volume, and Nicholas Philipson, the Editorial Director for Business, Economics, and Statistics at Springer Science + Business Media, as well as his diligent and hardworking team, for their initial and continual support of this project. Thanks to all our efforts and energy, this book has come to reality, and it will hopefully contribute to developing a contemporary framework of understanding and explanation of multilateral issues and variances of creativity, innovation, and entrepreneurship across cultures.

Barnaul, Russia Igor N. Dubina

Contents

1 **Culture as a Driving Force of Individual and Organizational Behavior** ... 1
Igor N. Dubina, Suzanna J. Ramos, and Hector Ramos

2 **Creativity Through a Cultural Lens: The Dichotomy of "The West" and "The East"** ... 29
Igor N. Dubina and Suzanna J. Ramos

3 **Wandering the World in Search of Creativity** ... 35
Robert Alan Black

4 **The Geography of the Creative Mind: A Cross-Cultural Study of Implicit Theories of Creativity Between the USA and Singapore** ... 51
Suzanna J. Ramos

5 **Group Creativity and Individual Creativity: A Case Study of the Differences Between Japanese and Chinese Creativity** ... 63
Fangqi Xu

6 **A Review on an Ancient Cultural Convergence: A Case Study on Arsacid Creativity and Inventiveness in Interrelationship with Their Greek Origin Ethnic Group** ... 77
Safura Borumand

7 **A Relationship Between Creativity, Management Styles, Culture, and Language (A Case of Japanese and English)** ... 87
Michihiko Esaki and Chris Nishihama

8 **Cross-Cultural Challenges for Innovation Management** ... 95
Maria Rosaria Della Peruta, Nigel J. Holden, and Manlio Del Giudice

9 **Cultural Influences on Innovation and Competitiveness** ... 109
Panagiotis E. Petrakis

10 Innovating Cultural Synergy in French–Russian Collaboration: Building a Puzzle from Cultural Diversity 121
Natalia I. Guseva

11 A Cultural Context, Curiosity, and Creativity of Innovators: Marie Curie, Nikola Tesla, and Steve Jobs 135
Phillip S. Harvard

12 Cultural Underpinnings in Entrepreneurship 147
Igor N. Dubina and Suzanna J. Ramos

Afterword 155

Index 157

Contributors

Robert Alan Black, Ph.D. S.P.R.E.A.D.ng Creative Thinking Throughout, Athens, GA, USA

Safura Borumand, Ph.D. History Department, Institute for Humanities and Cultural Studies, Kurdistan Highway, Tehran, Iran

Igor N. Dubina, Ph.D., Dr.Sc. Novosibirsk National Research State University, Novosibirsk, Russia

Altai State University, Barnaul, Russia

Michihiko Esaki, Ph.D. DTCN International, Gifu, Japan

Manlio Del Giudice, Ph.D. University of Rome "Link Campus", Rome, Italy

Natalia I. Guseva, Ph.D. National Research University Higher School of Economics, Moscow, Russia

Phillip S. Harvard, M.B.A., D.E.A., D.E.S.S. EIGSI General Engineering School, La Rochelle, France

Nigel J. Holden, Ph.D. Leeds University Business School, Leeds, UK

Chris Nishihama Gifu, Japan

Maria Rosaria Della Peruta, Ph.D. Second University of Naples, Naples, Italy

Panagiotis E. Petrakis, Ph.D. Department of Economics, University of Athens, Athens, Greece

Suzanna J. Ramos, Ph.D. Department of Educational Psychology, College of Education and Human Development, Texas A&M University, College Station, TX, USA

Hector Ramos, Ph.D. Department of Educational Psychology, College of Education and Human Development, Texas A&M University, College Station, TX, USA

Fangqi Xu, Ph.D. The Institute for Crerative Management and Innovation, Kindai University, Higashi-Osaka, Japan

Chapter 1
Culture as a Driving Force of Individual and Organizational Behavior

Igor N. Dubina, Suzanna J. Ramos, and Hector Ramos

Abstract Culture is deemed as a fundamental determinant of human behavior. Cross-cultural studies on management, business, and entrepreneurship practices represent a comparatively new class of interdisciplinary research. This movement represents the intersection of the cultural, sociological, psychological, historical, economic, management, technological, and business studies. This chapter encompasses the main definitions of culture and the theoretical assumptions of the impact of culture on creativity, innovation and entrepreneurship.

Introduction

Culture is deemed as a fundamental determinant of human behavior. Aristotle once said, "What is honored in a culture will be cultivated there" (as quoted in Raina 1999, p. 456). Culture underpins the "rules of the game" in any society and provides the informal constraints on human interaction (North 1990; Taeube 2004). In the context of the challenge of creativity research, Andrey Kolmogorov, a Russian academician, said, "One must model culture if one wants to model creativity" (as cited in Dubina 2000, p. 45).

Cross-cultural studies on management, business, and entrepreneurship practices represent a comparatively new class of interdisciplinary research. This movement, driven by both theoretical and practical interests, started 50 years ago. It represents the intersection of the cultural, sociological, psychological, historical, economic, management, technological, and business studies. Business studies, in particular, was connected with a growing number of large-scale multinational corporations and

I.N. Dubina, Ph.D., Dr.Sc. (✉)
Novosibirsk National Research State University, Novosibirsk, Russia

Altai State University, Barnaul, Russia
e-mail: igor_dubina@yahoo.com

S.J. Ramos, Ph.D. • H. Ramos, Ph.D.
Department of Educational Psychology, College of Education and Human Development, Texas A&M University, College Station, TX 77843, USA
e-mail: suzannaramos@gmail.com; hreina@gmail.com

I.N. Dubina, E.G. Carayannis (eds.), *Creativity, Innovation, and Entrepreneurship Across Cultures*, Innovation, Technology, and Knowledge Management, DOI 10.1007/978-1-4939-3261-0_1

the global context in which firms developed and implemented business strategies (Jones and Davis 2000). Thus learning culturally-appropriate business and innovation strategies became significant for managers (Shane 1995).

The book *The Chrysanthemum and the Sword* (1946) by Ruth Benedict, one of America's most famous anthropologists, can illustrate the importance of understanding cultural differences. In the second half of World War II, Benedict was assigned to study Japanese culture so that American forces could understand Japanese militarism and patriotism. She concretized the notion of the "East" and "West" dichotomy and the main patterns of Japanese culture, such as social hierarchy, honor, and duty. Her study was used by the US occupation administration in Japan from 1945 to 1952 to help the Americans understand and cooperate with the Japanese. This war-time research helped the Americans to interact with the Japanese and build a long lasting partnership.

An understanding of a culture can also help predict the behavior of typical members of the culture and explain some intangible factors by which all societies are governed. Cross-cultural studies discredited the myth of universalism in managerial thinking and practices (Amado et al. 1991). The development of management models is embedded in the cultural specificities of their environments, and it is therefore problematic to fully export them to other cultural contexts. The starting point for developing new and adjusting old managing practices in different geographical contexts should be driven by cultural values (García-Cabrera and García-Soto 2008). As Amado and colleagues (1991, p. 92) pointed out, the awareness, tolerance, and appreciation of cultural diversity help to succeed in establishing transcultural synergies in the modern age of internationalization and globalization and lead to flexible and innovative management approaches, an imperative for survival and development. Further, in an increasingly globalized international business environment, the effectiveness and efficiency of projects are affected by their cross-cultural characteristics (Chevrier 2003).

An extensive body of theoretical and empirical research on cross-cultural issues of general management describes diversity of entrepreneurship and managerial systems in various countries, cultural impact on managerial behavior, and other aspects of national cultures in the context of international business and management strategies (Andrews and Mead 2009; Bartlett and Ghoshal 1989; Chevrier 2003; Cullen and Parboteeah 2011; Ghoshal and Bartlett 1987; Hampden-Turner and Trompenaars 2000; Jones and Davis 2000; Moran et al. 2007; Mead 2005; Schneider and Barsoux 2003a, b; Smith et al. 2008; Thomas 2008; Usunier 1998). The management implications of identifying and understanding the differences in national cultures are wide ranging. A large literature (Earley and Erez 1997; Simons et al. 1993; Thiederman 1991) focuses on cultural diversities and appropriate behaviors in international work environment. Many authors are concerned with effective global leadership (House 2002; Simons et al. 1993), the relationship between culture and site selection when locating R&D (research and development) facilities overseas (Jones and Davis 2000), coordination and management in global teams (Harvey and Novicevic 2002), negotiation agreements and joint activities with potential foreign partners (Graham et al. 1994; Tung 1991), issues of global advertising (House et al. 2010) and other cross-cultural issues.

Cross-cultural interests in creativity, innovation, invention, and entrepreneurship have also resulted in wide-ranging publications. However, most cross-cultural research focus on highly developed countries like the United States, the European Union, Japan and fast-growing Asian economies such as China, Taiwan, South Korea, Singapore, and Indonesia (Begley and Tan 2001; Mitchell et al. 2002; Steensma et al. 2000a, b). Despite the increase of cross-cultural research many Latin American, Middle East, and African countries still remain unexplored (De Vreede et al. 1989; García-Cabrera and García-Soto 2008; Noorderhaven and Tidjani 2001).

A special perspective of cross-cultural research focuses on creativity. It examines how creativity is conceptualized differentially across cultures and how cultural factors account for differences in creativity. Research demonstrates that culture is the invisible force behind the tangibles and observables in any organization (e.g., Earley 1993, 1997; Ferraro 1998; Lubart 1990; Lubart and Georgsdottir 2004). There is sound evidence that culture may promote or inhibit creativity (Arieti 1976; Csikszentmihalyi 1988; Herbig 1994, 1996; Shane 1992b). Some cultures promote creativity more than others and they are known as "creativogenic cultures" (Arieti 1976). Empirical research has demonstrated a "cultural effect" in training creative skills, team performance, motivation, and satisfaction (Basadur 1992; Basadur et al. 1992; Li 1997). There is also evidence for the impact of culture on cognitive style and personality as related to creativity and innovation and the relationship between cultural values and creativity/innovation (Hoffman 1999; Hoffman and Hegarty 1993; Westwood and Low 2003). Some researchers (Basadur 1992; Basadur et al. 1992; Herbig 1996) have contrasted the creative/innovative achievements of Asian and Western countries, especially Japan and China with those of the United States and Europe. They have shown that, overall, culture can and does have an impact on the perception and interpretation of creative processes and cultural norms surrounding the process of both individual and group decision making.

There is also a considerable body of research that has examined innovation across cultures (nations). Recent research suggests a positive relationship between identified dimensions of culture and successful innovative activity. Researchers have studied the potential link between national culture, using national boundaries as a proxy for cultural boundaries, and the propensity to support innovative activities. This research stream has focused on the diffusion rate of innovations across borders (e.g., Ghoshal and Bartlett 1988), cultural differences in innovation processes and in innovative activity within organizations (Hoffman 1999; Hoffman and Hegarty 1993), the effect of national culture on preferences for championing strategies (Shane 1992b, 1995; Shane et al. 1995), the impact of national culture on R&D operations and investments (Jones and Davis 2000; Jones and Teegen 2001) and other issues.

An extension of innovation research focuses on the R&D globalization phenomenon (Jones and Davis 2000; Jones and Teegen 2001), national rates of innovation (Shane 1993), R&D initiation, implementation and productivity (Kedia et al. 1992; Nakata and Sivakumar 1996), national differences in the preference for licensing (Shane 1992a), adoption of innovations across cultures (Herbig 1994; Tellis et al.

2009), comparing the inputs of innovation, such as R&D spending, scientific personnel, and patents across nations (Archibugi and Coco 2005; Furman et al. 2002; Lin 2009) and national drivers of innovation (Archibugi and Coco 2005).

Numerous studies have linked national culture to the strategic decision-making process that occurs within entrepreneurial organizations (Kreiser et al. 2002). Many authors (e.g., Begley and Tan 2001) have stressed the importance of socio-cultural variables in explaining variations in entrepreneurship and economic development. Research interest has focused on understanding the influence of national cultural values on entrepreneurs and entrepreneurial actions (Mueller and Thomas 2001; Steensma et al. 2000a, b; Tiessen 1997).

Some research (Au 1997; García-Cabrera and García-Soto 2008) concludes that nationality and national cultures no longer play an important role because of the increasing economic and social globalization, arguing the need for new groups of variables to identify the preferences and attitudes of individuals in the business context (Zander and Romani 2004). Factors such as the development of technology and information systems, the globalization of markets and the increasing migratory flows across borders have affected the way in which organizations are designed and operated (Doz et al. 2001; Maznevski 1994). Sackmann and Phillips's (2004) argue that the national culture assumption is no longer valid in the current context, and a researcher should consider intra-cultural differences as well (García-Cabrera and García-Soto 2008).

Despite globalization, significant differences in cultures remain (Hofstede et al. 2004). Ravlin et al. (2000) argue that globalization does not address the problem of cross-cultural management. Moreover, globalization generates new difficulties with regard to managing workforce diversity, multicultural coordination, cooperation and communication.

The following sections of this chapter will encompass the main definitions of culture and the theoretical assumptions of the impact of culture on creativity, innovation and entrepreneurship.

The Concept of Culture: Problems of Definition, Delineation and Measurement

Although there are numerous definitions of culture, there is consensus that (1) culture affects people's minds, and (2) that there are many different aspects of this phenomenon (Kaasa and Vadi 2010). Aycan (2000) has proposed that the real issue is the extent and the ways culture influences individual and group phenomena in organizations.

In most definitions, culture is defined as values, rituals, and codes which refer to a core set or a system of cognitions, attitudes, behaviors, practices, customs, values, rules, and symbols. This core set is *shared* by the members of a collective entity, such as a nation or a firm and *governs* how people interact with their social and physical settings (Hofstede 1980, 2001; Lubart and Georgsdottir 2004; Tellis et al. 2009).

Culture also refers to a *learned* and *socially transmitted* set of behavioral standards. It is held, expressed, and shared by individuals through their personal values, norms, activities, attitudes, cognitive processes, interpretation of symbols, feelings, ideas, reactions and morals (Hofstede 1980, 2001; Morris et al. 1994). Culture is learned most intensively in the early years of one's life and has an enduring impact throughout life (Kaasa and Vadi 2010).

Therefore, culture can be defined as a learned set of norms for beliefs and social behavior that are shared by members of a group and which governs their behavior. These three main features: (1) culture is *shared* (2) culture is *transmitted* and (3) culture *governs,* establish a general context of culture delineations in which more specific definitions are suggested. For example, the three main features are embedded in the following definitions from the literature:

(a) "The shared motives, values, beliefs, identities, and interpretations or meanings of significant events that result from common experiences of members of collectives that are transmitted across generations" (House et al. 2001, p. 494–495).
(b) "The acquired knowledge that people use to interpret experience and to generate social behavior" (Rugman and Hodgetts 2003, p. 126).
(c) "That complex whole which includes knowledge, belief, art, morals, law, custom, and any other capabilities and habits acquired by man as a member of society"(see Kaasa and Vadi 2010, p. 6).
(d) "A collective mental knowledge developed by a group of people exposed to a similar context" (see Mitchell et al. 2002, p. 13).
(e) "A type of indicator of the optimal way of acting in the world and of understanding the world, and an indicator of the boundaries that influence the selection of experience in this optimal way" (Leontiev 2006, p. 52).
(f) "Collective programming of the mind which distinguishes the members of one group from another" (Hofstede 1980, p. 25).

Culture consists of several layers (Trompenaars and Hampden-Turner 1998): the core (assumptions about existence), middle layer (norms and values), outer layer (explicit products), which is most observable. According to Ulijin and Fayolle (2004), cultural layers could be presented as follows: basic assumptions (implicit), perceptions, attitudes, values and norms, artifacts and products (explicit).

People belong to many different "human groups" (a person may belong to a race, an ethnic group, a nation, a gender, a generation, and an organization all at the same time) and each of these groups has a culture that influences behavior (Hofstede 1980, 1991). Therefore , culture exists at multiple levels, ranging from broad societal or national cultures to group or corporate/ organizational cultures (see Mitchell et al. 2002). Three different but overlapping and interrelated contextual levels of culture are usually considered (Ulijn et al. 2000). They are (1) national culture, (2) corporate culture and (3) professional/occupational culture.

National culture studies are among the most intensely and widely examined and usually involve speculations and empirical research on how a country's national culture influences the behavior of its members and distinguish them from each other (Hofstede 1980, 2001).

The second prominent area of study has focused on *corporate culture,* or how members perceive the culture of their organization. Studies of this nature emphasize how the organization regulates, controls, and influences the behavior of its members through its values, language, rituals, and customs. National culture and organizational culture are composed of different elements. National culture consists of shared meanings and unconditional relationships, and its members are totally immersed, while organizational culture consists of changeable conditions and partly involves its members (Thomas 2008).

The third cultural level, and one less studied by business communication scholars, is *professional (occupational) culture.* Issues associated with cultural studies of this type include the extent to which professionals for example scientists, engineers, and managers, identify with their professional discipline rather than with their organization (Ulijn et al. 2001).[1]

While culture exists and can be studied at many levels, the purpose of this book is centered on the concept of national culture. So, we focus on this cultural aspect. As Jones and Davis (2000) pointed out, "Although there are multicultural nations and cultures that cross national boundaries, the concept of national culture—using national boundaries as a proxy for differentiating between distinct cultures and their respective cultural characteristics—has been generally accepted by researchers and practitioners. Similarities in national cultures are derived from common history, religion, geography and language, and even though there are differences within national borders, they are generally assumed to be of less significance than those differences found between nations."

The concept of national culture assumed in current research work corresponds to the common perspective seen as the set of values, beliefs and attitudes shared by the individuals of a human group, which influences individuals' behavior and social relationships, as well as a group's response to its environment and distinguishes one group from another this way (Hofstede 2001). Furthermore, House and his colleagues (2001, 2002, 2010) define national culture with the characteristics that distinguish cultures from each other and are predictive of organizational practices.

There are some common interactions that individuals face (see García-Cabrera and García-Soto 2008): the relationship with authority, the relationship between individual and society, individual concept of masculinity and femininity, and ways of dealing with conflicts including aggression control and expression of feelings. Assuming that the responses of individuals from the same nation coincide, cultures emerge as a result of the different responses that human groups offer to these basic dilemmas. These responses are given within the context of integrating forces such as national language, communications media, national army, and national sports teams (Hofstede 1980, 2001)

Hofstede's work (1980, 2001), considered a cornerstone in research on national cultures, makes a robust argument in favor of national culture. Nations are political entities, where factors such as legal, educational, political, and labor systems, as well as official languages, serve as mental programming for the members of a nation.

[1] Some issues of cultural levels are also discussed in Chap. 10 of this book.

Therefore, a nation is a social system which has its own culture. People often express their self-identities through their nations. Hofstede's model is widely analyzed in the literature, replicated and cited in numerous research works. It has been used successfully in numerous studies and taken as the foundation for developing subsequent models (a brief review see, e.g., in García-Cabrera and García-Soto 2008).

Despite many applications, the conception of national cultures developed and used by Hofstede and other researchers has been often questioned. Hofstede's main assumption is "one nation is one culture", so the nation is considered as a political unit, differentiating it from the concept of society (Kreiser et al. 2002; Dolan et al. 2004). However, geographical boundaries are not always the best criterion to study cultures (Cosmas and Sheth 1980). The "one nation—one culture axiom" assumes a single homogeneous national culture. This assumption can work for comparatively monochromic cultures such as Japan, but it may not be true for polychromatic cultures. For example, the Chinese do not represent a homogeneous group and the Chinese world is not static. Although the Chinese people from the People's Republic of China consist of Chinese from mainland China and Hong Kong and they share the same ethnicity, cultural roots, and social practices, the enormous historical and sociopolitical differences among them do not justify treating them as a uniform group (Rudowicz 2004). Therefore in the literature these cultures are always compared differently, although all of them are in the boundaries of the People's Republic of China.

Prior research has shown significant within-country differences in the levels of innovative activities, particularly in larger countries (Dakhli and de Clercq 2004). In such a context, various authors have explored multicultural differences in nations (e.g., see García-Cabrera and García-Soto 2008; Ardichvili and Gasparishvili 2001), sometimes with contrary conclusions. For example, Peppas (2001) stresses that the United States is a multicultural society. However, Kreiser et al. (2002) identify the United States as a distinctive monochromic national culture.

A statistical approach to culture using "cultural means" has been used to delineate cultures (Hofstede 1980, 2001; Au 1997). The more the difference in "cultural means", the more differently people see and describe each other. There are also some "deviations" associated with intra-cultural variations. A cross-cultural difference means that two or more groups of people are different from each other on one or more common features. If intra-cultural differences are greater than cross-cultural differences in means, the compared culture should be considered similar, since the difference in their "cultural means" is statistically insignificant. Consequently, the concept of national culture does not work in such a case. In contrast, this concept works if cross-cultural differences are greater than intra-cultural variances.

Thus, a researcher should assess the impact of intra-cultural variances when testing cultural mean differences. However, Hofstede's paradigm does not take into account variation within cultures although he recognized it himself. In his monumental work (1980, 2001) there are no standard deviations for single questions of the questionnaire. Scores (means) for responses are integrated, but variances are not indicated. In sum, it is convenient to use country boundaries as a proxy for the multitude of social, cultural, economic, and political differences. However, the concept of national culture may not be completely applicable because of intra-cultural differences.

In addition to the problems of heterogeneity of national culture and intra-cultural differences, there are several methodological problems in studying and measuring the cultural impact. This problem could be generally expressed in the form of the following questions:

1. What should we measure?
2. Whom should we measure?
3. How should we measure?

The first question, *what should we measure?* relates to the multi-impact effect. Human behavior is determined by many interrelated factors along with cultural determinants. Moreover, apart from national culture, many other sub-cultures (e.g. professional, organizational, gender) may influence individual behavior, making it challenging to delineate "pure" cultural determinants. Studies of cultural values often fail to control for differences in factors such as organizational structure, corporate culture, legal systems, technology, wealth, and economic systems (Shane 1993).

Some differences reported as 'cross-cultural' may have other reasons. For example, differences in creativity training across cultures (See, e.g., Basadur et al. 1992; Isaksen and Dorval 1993; Puccio and Gonzales 2004) may be explained by the differences in professional groups. In another example, the superior productivity of Japanese workers compared with their U.S. counterparts is commonly attributed to cultural reasons. However, Weiss (1984) asserts that such manifestations of behavior attributed to the inherent cultural proclivities of the Japanese worker as lower absenteeism, greater corporate loyalty, and harder working employees are myths. His suggested alternative "realities" of Japanese productivity are more engineers per worker, selective hiring, benefits from steep wage profiles, substantial pay differences, and a unique capital structure (Brannen 1991).

The second question, *whom should we measure?* mostly relates to a problem of a sample. As mentioned, along with national differences, the existence of intra-national differences is also increasingly emphasized (Au and Cheung 2004). It is not always possible to obtain all representative samples from large countries with multiple sub-cultures. Another aspect of this problem is the meaning of national culture for an individual who was born in one culture and lives in another. For example, comparisons of expatriates and people still living in their country of origin revealed no significant differences in preferences for championing strategies (Shane 1995). This suggests that one's culture of origin rather than one's host culture influences championing behavior. In fact, in a review of the international human resource management literature, Schneider (1988) argues that the dominant influence of culture of origin on organizational behavior exists because expatriates stress their culture of origin in reaction to the host culture.

The third question, *how should we measure?* mostly relates to the problem of indexes and measures. First, the examination of the effects of cultural values on other phenomena is problematic, since it is not easy to get accurate measures of cultural values, and to develop measures of culture that are reliable and valid that can be applied to various countries (Hambrick and Brandon 1988; Shane 1995). To measure a cultural impact on, for instance, creativity or innovation, it is challenging for the researcher to obtain reliable indexes as independent variables for such phenomena. For example, measuring innovation can be a tricky business (Tellis et al. 2009). Researchers who want to measure innovation often tend to focus on innovation inputs, such as per capita number of scientists, scientific articles or Nobel Prize winners, R&D expenditures, patents and trademarks, as proxies for innovation. However, these parameters could be a more appropriate measure of invention than for innovation as many ideas and patents never become viable products. Patents are usually used because they are easily measured, seem to be a precondition for innovation, or seem to offer protection to intellectual property. However, many high-tech firms now realize that patents provide only partial protection for their inventions, and firms can be highly innovative without patenting. Additionally, patenting intensity reflects not only the level of innovativeness or innovation initiation activity, but also the propensity to protect intellectual property (Kaasa and Vadi 2010).

Second, the problem of measuring cultural effects is inherent in methods of collecting and processing data. The existing approaches to measuring cultural values and other phenomena are based on questionnaires which use ordinal scales for coding respondents' replies (a Likert-type scale). However, in most cases, these scales are mistakenly interpreted as based on equal intervals. Researchers, therefore, incorrectly apply many statistical techniques for processing data and analyzing the obtained results. For example, the parametric methods frequently used for processing the results obtained with this scale are not appropriate for undertaking a correct quantitative analysis. In particular, the Pearson correlation coefficient is frequently used in cross-cultural studies without any additional grounds for an interval character of a used scale (Dubina 2006, 2009).

Therefore, culture is a phenomenon that is difficult to define, delineate or measure. As Hofstede (1993) notes, culture is a *construct*. This implies that culture is "not directly accessible to observation but inferable from verbal statements and other behaviors and useful in predicting still other observable and measurable verbal and nonverbal behavior". In essence culture remains to be a complex field of study.

Cultural Constitutes: Why and How Culture Matters

The main driving force of culture is its values. What people value and respect (e.g. diversity or uniformity, freedom or control, originality or conventionality) determines which behavior is appropriate. Culture influences an individual's choices. Runco (2004) explains: "Choices are a reflection of one's values and one's views of appropriate behavior. Individuals explore ideas which are consistent with (appropriate within) their culture. Individuals are socialized such that they acquire parental and cultural values. And these include standards for deciding when and how to express one's originality".

Although, gender, stage of life, socioeconomic position, and other factors may also play a role in defining what is "appropriate" for an individual, culture is considered to be primarily a moderator variable which affects the relationships among other variables (Hoffman 1999). Over time, the cultural values developed as a way of coping with environmental conditions become institutionalized through the use of rules, authority structures, and standard operating procedures (Meyer and Goes 1988). Consequently, when people establish organizations, the characteristics of these institutions reflect their cultural values. For instance, individual and organizational behavior reflects societal attitudes toward authority, trust, loyalty, commitment, motivation, control, discipline, communication, consultation, participation, coordination, and uncertainty (Tayeb 1988). While differences in organizational behavior have been found to influence rates of innovation, culturally-determined differences in these behaviors might explain national differences in rates of innovation (Shane 1993). Explanations as to why differences in cultural values may influence rates of innovation might be found in institutional theory. According to institutional theory, organizations are influenced by the societies in which they operate and exhibit their values. As organizational characteristics reflect societal values, managers might find that the organizational behaviors that promote innovation (identified in the management literature) are easiest to develop in certain societies, and these behaviors, in turn, might help to increase national rates of innovation (Shane 1993; Hofstede 2001)

Society and culture provide opportunities, supply a kind of reinforcement (reward) and model (demonstrate) creative (innovative, entrepreneurial) efforts. Culture can foster one type of activities and hinder another. For example, one's occupation may be a deep source of identity, providing social status and social recognition—the social identity of engineers, entrepreneurs, scholars, and the like is not the same for all countries. For a certain period in each culture there is a "cultural hero" (Estrin, 2009, p. 226) that serves as a model for creativity efforts. The idea of a 'cultural hero' is illustrated during the post-sputnik era where engineers and scientists were considered as one of most creative professions both in the United States and the USSR.

Cultural values may also affect individual traits, where people try to behave in ways consistent with their values (Schwartz and Sagiv 1995). For example, valuing conformity fosters compliant rather than unconventional behavior. Values serve as ideals or thoughts and hence, as guides for self-regulation. People may strive to reduce discrepancies they sense between their values and behavior by changing their behavior (Carver and Scheier 1981; Kluger and DeNisi 1996; Roccas et al. 2002).

Culture exists at multiple levels. The concept of multiple levels of sociocultural integration (Steward 1986) suggests that local cultures reflect particular ethnic, social, economic, ecological, and political complexities in which individuals are immersed. For example, in the area of entrepreneurship while we expect a common culture of entrepreneurship to exist, we also expect to observe within group differences: that the entrepreneurial sub-cultures within each country will also be congruent with the national culture. These differences are expected to be distinct, although perhaps observable only on closer examination.

Cultural values and practices affect what leaders and founders of businesses do (House et al. 2001). Founders of organizations are immersed in their own societal culture, and they are most likely to enact behavior patterns that are favored in that culture. Founders influence the behavior of sub-ordinate leaders and subsequent leaders by using discriminating management selection criteria, role modeling, and socialization. Further, the dominant cultural norms endorsed by societal cultures induce behavior patterns and organizational practices that are differentially expected and viewed as legitimate among cultures. Thus, the attributes and behaviors of leaders are, in part, a reflection of the organizational practices, which in turn are a reflection of societal cultures (House et al. 2001).

National culture has an effect on professional culture as well. Country-specific differences in managing R&D professionals abroad can be linked to the cultural dimensions identified by Hofstede (Shane 1992a, b). Hoppe (1993) notes that, "R&D professionals, despite their similarities, carry with them the (mostly invisible) norms of their country, as reflected in the country differences that exist for power distance, uncertainty avoidance, individualism, and masculinity. That is, they are similar in what they value at the workplace, but the degree to which they value it varies from country to country."

According to institutional theory, organizations conform to the prevailing institutional structure of the environment in which they operate (Granovetter 1985). The way in which firms take action, interpret action, and share that interpretation with others exhibits the values and norms of their societies and reflects the social order (see Shane 1993). Firms that do not behave in a way that achieves legitimacy die off, and those who survive develop the same approaches to solving business problems (Ibid).

Societal cultural values and practices also affect organizational culture and practices. Societal culture has a direct influence on organizational culture. The shared meaning from the dominant cultural values, beliefs, assumptions, and implicit motives endorsed by culture, results in common implicit leadership and organizational theories held by members of the culture (House 2002; House et al. 2004). Organizational culture cannot run counter to national culture (Mead 2005).

National culture and its sub-cultures have a motivational function in possessing "a directive force" that guides people's behavior (Ng and Smith 2004). A cultural motivational function can also be involved in creative, inventive and innovative behavior. It can be assumed that some societies (and their members) are more innovative than others because they possess certain cultural values. Such an assumption follows from Weber's discussion of the connection between fundamental religious ideas of Protestantism and economic behavior ("The Protestant Ethic and the Spirit of Capitalism"). Cultural values impact on work attitude, and an individual assimilates these values and attitudes from family, friends, school, and the cultural environment such as books and films. For example, if some of the cultural values are supportive of entrepreneurship, they may influence an individual to become an entrepreneur.

Culture has been found to be an important factor affecting management processes. In particular, cross-cultural studies show that certain managerial activities are more appropriate in some cultures than in others (Hofstede 1980, 1993; Hoffman 1999). For example, Amado et al. (1991) demonstrates the differences in management practices in American (functionalist) and French (personalist) organizations. In the functional and instrumental American vision, the company is perceived as a system of tasks to be accomplished, and the position of the actors in these structures is defined principally by their functions. It is a system of roles organized in a functional hierarchy of tasks to be accomplished, with the responsibility for these tasks assigned to various agents according to their competence. The manager's essential responsibility is to organize activities, coordinate tasks, and define responsibilities. On the other hand, in the French personalist and social model, the organization is first of all seen as a collectivity of persons to be managed, and the company is a system of persons organized in a social hierarchy. The conception of organizational structures reflects a need to differentiate degrees of authority and status of individuals. The key question on authority for a French manager is "Who has authority over whom?", and for an American manager it is, "Who is responsible for what?" Another example is the difference towards simplicity. American managers and professionals desire for simplicity. In Russia, for example, a too simple approach could produce skepticism as exemplified by a Russian proverb, "Simplicity is worse than theft".

Culture can encourage or inhibit emanating new ideas from the individuals or groups located in a certain culture. Creativity, inventiveness, and innovation are often hindered by problems that can be explained by tapping into concepts of culture. The role of culture is even mentioned as the first issue in the "Big 10" Innovation Killers presented by Wyckoff (2003). Culture unifies people's behavior, but it may also create barriers between them. People's beliefs and behaviors can contribute or block the process of developing and implementing new ideas. Culture is an appropriate concept to describe how innovation is influenced by various human factors. Culture affects innovation because it shapes the patterns dealing with novelty, individual initiatives and collective actions, and understandings and behaviors in regard to risks as well as opportunities (Kaasa and Vadi 2010).

Culture may constrain or encourage innovation, impact technology and innovation diffusion, and hinder new ideas. Culture evolves approved standards as well as deviations from those norms or innovations recognized by individuals and society. Therefore, some ideas can seem more "crazy" for one culture than for other, and consequently, more resisted. Inkster (2007) provides an example where "in the 'advanced' West, Darwinism may have been more resisted by intellectuals because of prior commitment to other intellectual and faith paths, but less resisted in Japan or China because there were no such prior commitments, and elements of traditional culture were actually conducive to acceptance. For example, the Shinto conception that there is no clear demarcation between inanimate objects or between humans and other creatures, or Buddhist notions that the quality of one's present life might determine the character of one's rebirth" (Inkster 2007).

Some cultures accept deviations more than others and there is a range of permissiveness across cultures. Cultures also vary on the extent to which they value perseverance, tolerance of ambiguity, and risk taking and other factors identified as important for creativity. Cultures may possess beliefs or attitudes that can foster or hinder creativity (Lubart 1990; Lubart and Georgsdottir 2004) such as cultural slogans like "Playfulness and fantasy are for children only "or "Success is good, and failure is bad".

However, there is no simple relationship between cultural characteristics and the propensity to create or innovate. There is a limit to the cultural explanations for variances in creative, innovative, entrepreneurial and other types of behavior in different social groups (Inkster 2007). In many instances of historical technological change, national cultures appear to have been of less importance than the environments for innovation. Societies can demonstrate an explosion of creative, inventive, innovative and other accomplishments in one sociopolitical and historical period and have a form of "creative recession" in another. Such booms and declines were observed throughout the stage of the development of both Western and non-Western civilizations (Rudowicz 2003, 2004). For example China has been experiencing an innovation boom for the last few decades, although Chinese cultural fundamentals have not changed much. If culture had been the only determining factor, China would have experienced innovation sooner.

The complex and non-linear character of cultural influence on creativity and other phenomena have produced some controversial conclusions in the literature. For example, cultural collectivism is usually considered as an inhibitor of creativity development. But such a collectivistic society like China with a high respect of tradition has been respected across the globe for unique arts and inventions such as paper, printing techniques, the seismograph, silk, gunpowder, and the compass. Currently China is one of the leaders in the innovation technological market. Further, a collectivistic society like Japan is often deemed as not an inventive society as Japan does not produce the type of brilliant inventions found in the United States. However Japan is also one of the most important leaders in the innovation technological market. For instance, the Japanese car industry led by brands such as Toyota, provides a greater technological offer in terms of car reliability and quality to the world market than competitors elsewhere (Getz and Robinson 2003).

In Asian cultures, educators are greatly respected, and it would be most often clearly inappropriate to question or offer ideas that differ from the teacher. As some researchers (Ng 2001; Ng and Smith 2004) argue, it could restrain creativity. On the other hand, certain Asian institutions are very appreciative to original ideas such as the use of suggestion boxes and other similar techniques in Japanese organizations (Basadur 1992).

Asian cultures are often considered less risk-taking than "western" societies because of the "shame effect" or "losing-of-face effect". Lubart and Georgsdottir (2004) refer to a study examining proverbs commonly used in the United States and China concerning risk-taking. Contrary to popular belief, risk-taking was considered more negative in the financial domains according to American proverbs than the Chinese ones. This result could be explained by the "cushion" effect, where the collectivist nature of Chinese culture reduces the negative impact of financial risk-taking because the family and community can 'cushion' failure.

These and many other controversial opinions and findings do not confirm the absence of cultural impact on creativity and innovation. They just confirm that (a) the influence of culture is very complex and nonlinear, (b) cultural impact interplays with historical, political, economics, and other factors. Presently, no conclusive and comprehensive theory of this relationship has been offered yet.

Cultural Dimensions: How Cultures Can Be Measured and Classified

Several taxonomies identifying the main features or *cultural dimensions* have been developed to research the cultural impact on different aspects of social and psychological life. A considerable body of research deals with cultural peculiarities and dimensions to explain why and how culture manifests in human practices. These include: *High- and Low context cultures, Individualism/Collectivism, Assertiveness/Responsiveness, Idiocentric/Allocentric, Self-enhancement/Self-transcendence, Openness to Change/Conservation, Self-Direction/Universalism, Individual Dynamics/Group Dynamics, Hedonism, Tradition, Conformity, Communication styles,* and *Uncertainty Avoidance* (see reviews in Schwartz and Sagiv 1995; Hofstede 2001; Thomas 2008; some more details on cultural dimensions are also provided in Chap. 10 of this monograph).

Hofstede's Model of Cultural Dimensions

Hofstede's model of cultural dimensions is the most well-known approach to study cultural impact. This model serves as the basis for the majority of the theoretical and empirical research over the past four decades on most business and management aspects of national culture. Hofstede's work (1980, 1991, 2001) has become the quintessential representation of how national cultures influence business issues. His cultural dimensions are constructs which "do not exist; they are tools for analysis which may or may not clarify a situation" (Hofstede 1993).

The following dimensions were initially detected through a comparison of the values of more than 100,000 employees and managers in 64 national subsidiaries of the IBM Corporation. People working in different countries represented samples from the populations of their countries, similar in all respects except nationality. Hofstede (1980) identified the key cultural characteristics which appear to distinguish among cultures:

1. Power Distance: the tolerance of social inequality, or the degree to which people accept authority and status differences in society and their organizations.
2. Uncertainty Avoidance: intolerance of ambiguity and uncertainty, or the extent to which people fear ambiguous situations and seek to avoid them.
3. Individualism versus Collectivism: the preference for behavior that promotes one's self-interest, or a people's self-concept of being independent actors or being dependent on a group.
4. Masculinity versus Femininity: the extent to which people value masculine values of assertiveness, materialist and achievement orientation or feminine values of cooperation and aesthetics.

These values deal with four different anthropological problem areas: ways of coping with inequality, ways of coping with uncertainty, the relationship of the individual with her or his primary group, and the emotional implications of having been born as a girl or as a boy.

Eventually, a fifth dimension of national cultures was added. It was based on a study of the values of students in 23 countries (with Michael Harris Bond). It was labeled first as Confucian Dynamism, since it basically reflected the difference between a dynamic, future-oriented society (positive Confucian dynamism, longer-term perspective) versus a more static, tradition-oriented one (negative Confucian dynamism, shorter term perspective). This dimension was later labeled as Long-term versus Short-term Orientation (Hofstede 1993).

Here is a summary of Hofstede's cultural dimensions.

(a) Power Distance
Low-power distance cultures do not accept that power be distributed unequally, in contrast to countries with a high score in this dimension, which consider inequality to be an inherent feature of society. These high-power distance cultures accept and expect more powerful individuals to possess certain privileges. In low power-distance cultures forces toward centralization of decision making are expected to be weaker with a consequent lower relationship between such forces and decentralization and delegation practices (House et al. 2001). On the one hand, in high-power distance cultures, hierarchical differences are valued (i.e. powerful people are expected to display their status and all expect them to claim special privileges). On the other hand, in low-power distance cultures, equality is valued and those with more power or status should not act in ways that call attention to their advantages.

(b) Uncertainty Avoidance
Strong uncertainty avoiding countries typically feel threatened by ambiguous situations and design ways to reduce their stress and fear of the unknown. The stress resulting from uncertainty leads to the need to adopt formal rules, a greater preference for bureaucratic organizations, as well as less tolerance of people or ideas that diverge from already familiar models. Individuals fear failure in these societies and consequently, their members avoid taking risks. Low-uncertainty avoidance societies fully accept uncertainty. Such countries exhibit a higher level of tolerance for change, ambiguity, and accept and often embrace the risks associated with an uncertain future. In low-uncertainty avoidance cultures, forces toward formalization should be weaker, and therefore the relationship between such forces and organizational formalization practices will be lower (House et al. 2001). In societies with low-uncertainty avoidance, organizational rules can be violated for pragmatic reasons. Conflicts are considered a natural part of life, and ambiguous situations are regarded as commonplace and appealing. In the case of strong uncertainty avoidance, the opposite tends to prevail. In working relations, rules play an important role and are carefully followed. Cultures with low uncertainty avoidance tend to meet basic demands, are 'tolerant of various behaviors and feel relatively secure' (Nakata and Sivakumar 1996) and are 'more prepared to give the benefit of the doubt to unknown situations, people, and ideas' (Hofstede 1991).

(c) Individualism versus Collectivism
Individualistic cultures value self-orientation, self-sufficiency and self-control, the pursuit of individual goals that may or may not be consistent with in-group goals, a willingness to confront members of the in-group to which a person

belongs, and people's own accomplishments. In an individualistic cultural environment, people are motivated by self-interest and achievement of personal goals. They are hesitant to contribute to collective action unless their own efforts are recognized, preferring instead to benefit from the efforts of others (Morris et al. 1994). Collectivistic cultures focus on the subordination of personal interests to the goals of the larger work group: an emphasis on sharing, cooperation, group harmony, a concern with group welfare, and hostility toward out-group members. In a collectivistic environment, people feel personally responsible for the group product and are oriented towards sharing group rewards.

(d) Masculinity versus Femininity
Masculine societies are aggressive, task and performance oriented, and include many occupations that are typically considered gender-specific. A masculine society favors challenge, advancement and the accumulation of money. The main characteristics of a masculine society are ambition, the need to excel, a tendency to polarize, admiration for the achiever, and decisiveness, the principles being "live in order to work" and "big and fast are beautiful" (Hofstede 1980, 1984, 1991, 2001). In feminine societies economic growth may not necessarily be the primary concern of the society. They are characterized by an emphasis on relationships and social interactions, a friendly atmosphere, job security, serving others, sympathy for the unfortunate, and striving for consensus and cooperation. The main principles of a feminine society are "work in order to live" and "small and slow are beautiful" (Hofstede 1980, 1984, 1991, 2001).

(e) Long-term versus Short-term Orientation
In long-term oriented cultures, the main work values are learning, honesty, adaptiveness, accountability, and self-discipline. In this culture people invest in life-long personal networks and leisure time is not valued. Instead, the focus is on market position where the owner/managers and workers share the same aspirations. In short-term-oriented cultures, the main work values are freedom, rights, achievement, and thinking for oneself. Personal loyalties vary with business needs. The focus is on the profits of the current year or quarter, and managers and workers are psychologically in two different camps. Long- and short-term oriented cultures represent two different ways of thinking, which can be characterized with the opposing labels 'virtue' versus 'truth' or 'synthetic' versus 'analytical' (Hofstede and Minkov 2010).

In the 2010 edition of *Cultures and Organizations* (Hofstede et al. 2010), a sixth dimension was added, based on Michael Minkov's analysis of the World Values Survey data for 93 countries. This new dimension was called Indulgence versus Restraint (the extent to which members of a society try to control their desires and impulses). Hofstede's dimensions scores for many national cultures (countries) are available in his works (1980, 2001) and can be found online at www.geert-hofstede.com and http://www.geerthofstede.nl

The five/six dimension cultural framework developed by Hofstede has been applied extensively by researchers in many fields. The validity and reliability of these measures have been demonstrated in numerous studies (see reviews in

Hofstede 2001). The validity of Hofstede's indices to people outside IBM can also be seen from correlations between his indices and those of other researchers. Indices such as personal and interpersonal values, capacity leadership, sharing information and participation, openness versus secrecy, attitudes toward older and younger people and occupational inheritance correlate significantly with at least one of Hofstede's dimensions (Shane 1992a, 1993, 1995; Lin 2009). Some researchers discovered similar dimensions in completely different material including different samples, questionnaires and scales. The findings represented sound support for Hoftstede's cultural dimensions (Hofstede and Bond 1988; Hofstede 1993).

Although Hofstede's conceptualization of culture is the most widely adopted, it is not without its critics (see, e.g., Gannon and Audia 2000; García-Cabrera and García-Soto 2008; Kaasa and Vadi 2010; Mead 2005; Thomas 2008; van Everdingen and Waarts 2003). The main criticisms are as follows:

- The sample was not randomly selected from the population, and represented only those people who worked for IBM.
- Although culture does not change rapidly as Hofstede (2001) argues, the collection of data may be out of date.
- Multiformity of the dimensions / factors (e.g., different types and sub-factors of individualism and collectivism are found (Triandis 1995, 1996, 2002).
- Possible biases in the questionnaire responses e.g. some questionnaires were completed not individually, but in groups in IBM subsidiaries.
- Methodological and technical issues (e.g. a few points in the scale were used; many items within dimensions relate to several of the dimensions).
- The acceptance of the nation as a suitable unit for analyzing cultures as it underestimates the importance of sub-cultural differences or differences between individuals within the same country.

In addition, we would also like to indicate a non-independent structure of Hofstede's measures that may produce a multicollinearity effect. There is a strong correlation between two of the dimensions (Table 1.1). Generally speaking, this makes it impossible to include the corresponding variables representing these dimensions in a regression model, although many researchers commit such an error.

Table 1.1 Correlation matrix for Hofstede's measures

	PDI	IDV	MAS	UAI	LTO
PDI	1				
IDV	−0.604	1			
MAS	0.078	0.110	1		
UAI	0.183	−0.187	−0.036	1	
LTO	0.279	−0.407	0.111	−0.077	1

Source: Calculations made by the authors with use data from (Hofstede 2001)

In response to this criticism, Hofstede updated and continued updating his measures for several countries using different samples representing broader societies (see Hofstede's answer to criticism (2001) and updated scores at www.geert-hofstede.com).

Despite its limitations, this model is presently the most useful in terms of number of cultures and respondents, reliability and applicability (Mead 2005; Andrews and Mead 2009). Hofstede's model has become a dominant paradigm of the "normal science" in cross-cultural studies (in term of T. Kuhn (1962)). While these weaknesses do not invalidate the usefulness of Hofstede's measures, the reader should keep in mind the limitations of the measures of cultural values employed here.

Other Models of Cultural Dimensions

The GLOBE (Global Leadership and Organizational Behavior Effectiveness) research program studies the effectiveness of leadership behaviors in different cultural contexts. The GLOBE team of cross-cultural researchers collected and analyzed data from approximately 17,000 managers from 951 organizations in 62 societies throughout the world (House et al. 2004). The GLOBE was designed initially to provide a solid foundation for global advertising research and managing global (international) teams (House, et al. 2001). The meta-goal of the GLOBE was to develop an empirically-based theory to describe, understand, and predict the impact of specific cultural variables on leadership and organizational processes and their effectiveness.

The GLOBE study reported nine constructs or dimensions: *performance orientation*, *assertiveness*, *future orientation*, *humane orientation*, *institutional collectivism*, *in-group collectivism*, *gender egalitarianism*, *power distance*, and *uncertainty avoidance* (House et al. 2004). Original scales were developed for each of these dimensions to reflect both the practices and values associated with each dimension within a given culture. As a result, 18 scales measured the practices and values associated with the nine core GLOBE dimensions of culture (detailed descriptions and analysis are provided in Javidan and House 2001; House et al. 2004, 2006, 2010; Javidan, Dorfman et al. 2006; Javidan, House et al. 2006).

Performance Orientation. The degree to which an organization or society encourages and rewards (and should encourage and reward) group members for performance improvement and excellence. This dimension includes the future-oriented component of the dimension called Confucian Dynamism by Hofstede and Bond (1988). In countries like the United States and Singapore that score high on this cultural practice, businesses are likely to emphasize training and development. Examples of countries with low score are Russia and Greece.

Assertiveness. The degree to which individuals (members of a society) are (and should be) assertive, confrontational, and aggressive in their relationships with others. People in highly assertive countries such as the United States and Austria tend to have can-do attitudes and enjoy competition in business. Those in less assertive countries such as Sweden and New Zealand prefer harmony in relationships and emphasize loyalty and solidarity.

Future Orientation. The extent to which individuals in organizations or societies engage (and should engage) in future-oriented behaviors such as delaying gratification, planning, and investing in the future. Organizations in countries with high future-oriented practices like Singapore and Switzerland tend to have longer term horizons and more systematic planning processes, but they tend to be averse to risk-taking and opportunistic decision making. In contrast, corporations in the least future-oriented countries like Russia and Argentina tend to be less systematic and more opportunistic in their actions.

Humane Orientation. The degree to which a collective encourages and rewards (and should encourage and reward) individuals for being fair, altruistic, generous, caring, and kind to others. This dimension is similar to the dimension labeled Kind Heartedness by Hofstede and Bond (1988). Countries like Egypt and Malaysia rank very high while countries like France and Germany rank low on this cultural practice.

Institutional Collectivism. The degree to which organizational and societal institutional practices encourage and reward (and should encourage and reward) collective distribution of resources and collective action. Organizations in collectivistic countries like Singapore and Sweden tend to emphasize group performance and rewards, whereas those in the more individualistic countries like Greece and Brazil tend to emphasize individual achievement and rewards.

In-Group Collectivism. The degree to which individuals express (and should express) pride, loyalty, and cohesiveness in their organizations or families. Societies like Egypt and Russia take pride in their families and also take pride in the organizations that employ them.

Gender Egalitarianism. The degree to which a society minimizes (and should minimize) gender role differences and gender discrimination while promoting gender equality. European countries generally had the highest scores on gender egalitarianism practices. Egypt and South Korea were among the most male-dominated societies in GLOBE. Organizations operating in gender egalitarian societies tend to encourage tolerance for diversity of ideas and individuals.

Power Distance. The degree to which members of a society expect (and should expect) power to be distributed equally, or the extent to which individuals agree that power should be unequally shared, i.e. stratified and concentrated at higher levels of an organization or government. A high power distance score reflects unequal power distribution in a society. Countries that scored high on this cultural practice are more stratified economically, socially, and politically. Those in positions of authority expect, and receive, obedience. Firms in high power distance countries like Thailand, Brazil, and France tend to have hierarchical decision-making processes with limited one-way participation and communication.

Uncertainty Avoidance. The extent to which a society, organization, or group relies (and should rely) on social norms, rules, and procedures to alleviate unpredictability of future events. The extent to which members of a society seek certainty and strive to avoid uncertainty in their environment by relying on established social norms, rituals, and bureaucratic practices to alleviate the unpredictability of future events. The greater the desire to avoid uncertainty, the more people seek orderliness, consistency, structure, formal procedures, and laws to address situations in their daily lives. Organizations in high-uncertainty avoidance countries like Singapore and Switzerland tend to establish elaborate processes and procedures and prefer formal detailed strategies. In contrast, firms in low uncertainty avoidance countries like Russia and Greece tend to prefer simple processes and broadly-stated strategies. They are also opportunistic and enjoy risk-taking.

Although based on a smaller sample, GLOBE is more current as the GLOBE study was conducted some twenty years after the Hofstede study. Another important dimension of the GLOBE study is that it attempts to capture both societal cultural norms of shared values in society, (the 'should be' values), as well as how they are practiced in society (the 'as is' values) (House et al. 2004).

Despite the similarity with Hofstede's dimensions there are some differences. For example, contrary to Hofstede's findings, Russia and Greece both reported low scores on uncertainty avoidance in the GLOBE study. These countries tend to prefer simple processes and broad strategies, leaving room for flexibility and risk-taking (Javidan, Dorfman et al. 2006; Javidan, House et al. 2006). The criticism of the GLOBE project measures, as well as the authors' responses to this criticism, is presented in (House et al. 2006).

One more taxonomy of cultural dimensions is based on the value theory developed by Schwartz (Schwartz and Sagiv 1995). It defines values as desirable, trans-situational goals that vary in their importance as guiding principles in people's lives. Schwartz and colleagues derived ten types of values, each of which expresses a distinct motivational goal: self-direction, tradition, conformity, universalism, benevolence, security, power, achievement, hedonism, and stimulation. They also specified the structure of relations among these values. The model has been tested in more than 200 samples from over 60 countries (Roccas et al. 2002).

Self-direction values emphasize autonomy of thought and action, openness to change, curiosity, and creativity. They call for self-exploration and independent judgment of people and events. These values are relevant to and compatible with independence of thought and self-reliance in making a career decision.

Tradition values emphasize submission and commitment to prevailing beliefs, practices and institutions rather than seeking out and adopting new ways of believing, behaving, and being. Independent behavior defies the unquestioned acceptance of authority and prevailing modes of action inherent in tradition values.

Conformity values emphasize restraint of actions or inclinations that might violate social expectations or norms or upset others. Like tradition values, conformity values promote maintenance of the status quo.

Universalism values emphasize understanding, appreciation, tolerance, and concern for the welfare of all people and for nature. The sub-type of universalism values, the *social-concern,* emphasizes care, concern, and protection for the welfare of other people, including those outside one's own circle. With their focus on the welfare of others, they are largely irrelevant to behavior toward the counselor or toward the pursuit of insights regarding self.

Benevolence values emphasize concern and care for others with whom one has frequent contact, that is, members of one's in-groups.

Security values represent the motivation for stability, harmony and security for society and close others and self.

Power values represent the motivation to dominate and control people and resources.

Achievement values emphasize personal success through demonstrating competence according to social standards.

Hedonism values emphasize pleasure and sensual gratification.

Stimulation values emphasize the pursuit of affective arousal through novelty, change, and excitement.

This model reflects the conflict and compatibility among values that people experience as they pursue different value priorities. Values that share compatible motivational goals correlate most positively. Values that express conflicting motivational goals correlate less positively, or even negatively. Each value primarily expresses a particular motivational goal (e.g. power, security, stimulation). The full set of values presumably represents the range of distinct human motivations (Schwartz and Sagiv 1995; Sagiv and Schwartz 2004; Roccas et al. 2002).

This model seems to be biased since it exaggerates mostly "western" values (e.g. originality, independence). The model has not been widely applied unlike Hofstede's model and it has received scant attention in the literature.

Conclusion

Culture is a fuzzy, multi-faceted and complex construct that impacts individuals' values, norms, and behaviors on multiple levels. In particular, culture may raise definitional, conceptual, and operational obstacles for research on itself and on its impact on individual and organizational creativity, as well as innovation and entrepreneurship. However, through our examination of the various models of cultural dimensions that drive organizational behavior, determining the influence of culture on organizational behavior remains a challenging task. This is because culture is a process in itself and the manifestations of culture often occur at a sub-conscious level (Hofstede 1991).

Despite the challenge of incorporating culture in understanding organizational behavior, there has been an increase in studies on culture in the business arena over the last three decades. The workplace is typically the epicenter of such research—a complex social system consisting of multiple levels of human activity. Researchers, especially cross-cultural and organizational psychologists, have established the significance of studying how culture influences various social systems. Models such as Hoftstede's and the GLOBE can continue to pave the way for future research on how culture drives individual and organizational behavior, especially in the area of creativity, innovation and entrepreneurship.

References

Amado, G., Faucheux, C., & Laurent, A. (1991). Organizational change and cultural realities: Franco-American contrasts. *International Studies of Management & Organisation, 21*(3), 62–69.

Andrews, T. G., & Mead, R. (2009). *Cross-cultural management: Critical perspectives on business and management*. New York: Routledge.

Archibugi, D., & Coco, A. (2005). Measuring technological capabilities at the country level: a survey and a menu for choice. *Research Policy, 34*(2), 175–194.

Ardichvili, A., & Gasparishvili, A. (2001). Socio-cultural values, internal work culture and leadership styles in four post-communist countries. Russia, Georgia, Kazakhstan and the Kyrgyz Republic. *International Journal of Cross Cultural Management, 1*, 227–242.

Arieti, S. (1976). *Creativity: The magic synthesis*. New York: Basic Books.

Au, K. (1997). Another consequence of culture - intra-cultural variation. *The International Journal of Human Resource Management, 8*(5), 743–755.

Au, K., & Cheung, M. W. (2004). Intra-cultural variation and job autonomy in 42 countries. *Organization Studies, 25*, 1339–1362.

Aycan, Z. (2000). Cross-cultural industrial and organizational psychology. contributions, past developments, and future directions. *Journal of Cross-Cultural Psychology, 31*, 116–128.

Bartlett, C. A., & Ghoshal, S. (1989). *Managing across borders: The transnational solution*. Boston: Harvard Business School Press.

Basadur, M. (1992). Managing creativity: a Japanese model. *Academy of Management Executive, 6*(2), 29–42.

Basadur, M. S., Wakabayashi, M., & Takai, J. (1992). Training effects on the divergent thinking attitudes of Japanese managers. *International Journal of Intercultural Relations, 16*, 329–345.

Begley, T. M., & Tan, W.-L. (2001). The socio-cultural environment for entrepreneurship: a comparison between East Asian and Anglo-Saxon countries. *Journal of International Business Studies, 32*(3), 537–553.

Brannen, M. Y. (1991). Culture as the critical factor in implementing innovation. *Business Horizons, 34*(6), 59–67.

Carver, C. S., & Scheier, M. F. (1981). *Attention and self-regulation: A control-theory approach to human behavior*. New York: Springer.

Chevrier, S. (2003). Cross-cultural management in multinational project groups. *Journal of World Business, 38*(2), 141–149.

Cosmas, S. C., & Sheth, J. N. (1980). Identification of opinion leaders across cultures: an assessment for use in the diffusion of innovation and ideas. *Journal of International Business Studies, 11*, 66–73.

Csikszentmihalyi, M. (1988). Society, culture, person: A systems view of creativity. In R. J. Sternberg (Ed.), *The nature of creativity* (pp. 325–339). Cambridge: Cambridge University Press.

Cullen, J. B., & Parboteeah, K. P. (2011). *Multinational management: A strategic approach*. Mason, OH: CL-South-western.

Dakhli, M., & de Clercq, D. (2004). Human capital, social capital, and innovation: a multi-country study. *Entrepreneurship & Regional Development, 16*, 107–128.

De Vreede, G.-J., Jones, N., & Mgaya, R. J. (1989). exploring the application and acceptance of group support systems in Africa. *Journal of Management Information Systems, 15*(3), 197–234.

Dolan, S. L., Martırn-Prius, A., Dırez-Piñol, M., Martırnez-Fierro, S., & Fernarndez-Alles, M. (2004). Exploratory study of within-country differences in work and life values. *International Journal of Cross Cultural Management, 4*, 157–180.

Doz, Y., Santos, J., & Williamson, P. (2001). *From global to metanational*. Boston: Harvard Business School Press.

Dubina, I. N. (2000). *Creativity as a phenomenon of social communication*. Novosibirsk: SO RAN (in Russian).

Dubina, I. N. (2006). Measurements in creativity management: why and how to measure intangibles. *International Journal of Knowledge, Culture and Change Management, 6*(6), 149–155.

Dubina, I. N. (2009). *Managing employees' creativity in innovation economy*. Moscow: Academia (in Russian).

Earley, P. C. (1993). East meets West meets Mid-east: further explorations of collectivistic and individualistic work groups. *Academy of Management Journal, 36*, 319–348.

Earley, P. C. (1997). *Face, harmony, and social structure: An analysis of organizational behavior across cultures*. New York: Oxford University Press.

Earley, P. C., & Erez, M. (1997). *The transplanted executive*. Oxford: Oxford University Press.

Estrin, J. (2009) Closing the innovation gap : reigniting the spark of creativity in a global economy. New York: McGraw-Hill.

Ferraro, G. P. (1998). *The cultural dimension of international business*. Upper Saddle River, NJ: Prentice Hall.

Furman, J. L., Porter, M. E., & Stern, S. (2002). The determinants of national innovative capacity. *Research Policy, 31*(6), 899–933.

Gannon, M. J., & Audia, P. G. (2000). The cultural metaphor: a grounded method for analyzing national cultures. In P. C. Earley & H. Singh (Eds.), *Innovations in international and cross-cultural management* (pp. 91–106). Thousand Oaks, CA: Sage Publications.

García-Cabrera, A. M., & García-Soto, M. G. (2008). Cultural differences and entrepreneurial behaviour: an intra-country cross-cultural analysis in Cape Verde. *Entrepreneurship & Regional Development, 20*, 451–483.

Getz, I., & Robinson, A. G. (2003). Innovate or die: is that a fact? *Creativity and Innovation Management, 12*(3), 130–136.

Ghoshal, S., & Bartlett, C. (1987). Management across borders: new strategic requirements. *Sloan Management Review, 43*, 1–17.

Ghoshal, S., & Bartlett, C. (1988). Creation, adoption, and diffusion of innovation by subsidiaries of multinational corporations. *Journal of International Business Studies, 19*(3), 365–388.

Graham, J. L., Mintu, A. T., & Rodgers, W. (1994). Exploration of negotiation behaviors in ten foreign cultures using a model developed in the United States. *Management Science, 40*(1), 72–95.

Granovetter, M. (1985). Economic action and social structure: the problem of embeddedness. *American Journal of Sociology, 91*, 481–510.

Hambrick, D., & Brandon, G. (1988). Executive values. In D. Hambrick (Ed.), *The executive effect: Concepts and methods for studying top managers*. Greenwich, CT: JAI Press.

Hampden-Turner, C. M., & Trompenaars, F. (2000). *Building cross-cultural competence: How to create wealth from conflicting values*. New Haven, CT: Yale University Press.

Harvey, M., & Novicevic, M. M. (2002). The co-ordination of strategic initiatives within global organizations: the role of global teams. *International Journal of Human Resource Management, 13*(4), 660–676.

Herbig, R. (1994). *The innovation matrix: Culture and structure, prerequisites to innovation*. Westport, CT: Quorum Books.

Herbig, P. (1996). Creative problem-solving styles in the USA and Japan. *International Marketing Review, 13*(2), 63–71.

Hoffman, R. C. (1999). Organizational innovation: management influence across cultures. *Multinational Business Review, 7*(1), 37–49.

Hoffman, R. C., & Hegarty, W. H. (1993). Top management influence on innovations: effects of executive characteristics and social culture. *Journal of Management, 19*(3), 549–574.

Hofstede, G. (1980). *Culture's consequences: International differences in work-related values*. Beverly Hills, CA: Sage Publications.

Hofstede, G. (1984). *Culture's consequences: International differences in work-related values*. Beverly Hills, CA: Sage Publications.

Hofstede, G. (1991). *Cultures and organizations: Software of the mind*. London: McGraw Hill.

Hofstede, G. (1993). Cultural constraints in management theories. *Academy of Management Executive, 7*(1), 81–94.

Hofstede, G. (2001). *Culture's consequences: Comparing values, behaviors, institutions, and organizations across nations*. Thousand Oaks, CA: Sage Publications.

Hofstede, G., & Bond, M. (1988). The Confucius connection: from cultural roots to economic growth. *Organizational Dynamics, 16*, 5–21.

Hofstede, G., Hofstede, G. J., & Minkov, M. (2010). *Culture and organizations - Software of the mind: Intercultural cooperation and its importance*. New York: McGraw-Hill.

Hofstede, G., & Minkov, M. (2010). Long- versus short-term orientation: new perspectives. *Asia Pacific Business Review, 16*(4), 493–504.

Hofstede, G., Noorderhaven, N. G., Thurik, A. R., Uhlaner, L. M., Wennekers, A. R. M., & Wildeman, R. E. (2004). Culture's role in entrepreneurship: self-employment out of dissatisfaction. In T. E. Brown & J. Ulijn (Eds.), *Innovation, entrepreneurship and culture: The interaction between technology, progress and economic growth* (pp. 162–203). Northampton, MA: E. Elgar Pub.

Hoppe, M. H. (1993). The effect of national culture on the theory of practice of managing R&D professionals abroad. *R&D Management, 23*(4), 313–325.

House, R. (2002). Understanding cultures and implicit leadership theories across the globe: an introduction to project GLOBE. *Journal of World Business, 37*(1), 3–10.

House, R., Hanges, P., Javidan, M., Dorfman, P., & Gupta, V. (2004). *Culture, leadership and organisations – The GLOBE study of 62 societies*. Thousand Oaks, CA: Sage Publications.

House, R. J., Javidan, M., Dorfman, P. W., & de Luque, M. S. (2006). A failure of scholarship: response to George Graen's critique of GLOBE. *Academy of Management Perspectives, 20*(4), 102–114.

House, R., Javidan, M., & Dorfman, P. (2001). Project GLOBE: an Introduction. *Applied Psychology: An International Review, 50*(4), 489–505.

House, R. J., Quigley, N. R., & de Luque, M. S. (2010). Insights from project GLOBE. *International Journal of Advertising, 29*(1), 111–139.

Inkster, I. (2007). Technology in world history: cultures of constraint and innovation, emulation, and technology transfers. *Comparative Technology Transfer and Society, 5*(2), 108–127.

Isaksen, S. G., & Dorval, K. B. (1993). An inquiry into cross-cultural creativity training: results from a five-week study tour in Bergen and Bratislava. In S. S. Grryskiewicz (Ed.), *Discovering creativity: Proceedings from the 1992 creativity and innovation conference* (pp. 151–154). Greensboro, NC: Center for Creative Leadership.

Javidan, M., Dorfman, P. W., de Luque, M. S., & House, R. J. (2006). In the eye of the beholder: cross cultural lessons in leadership from project GLOBE. *Academy of Management Perspectives, 20*, 67–90.

Javidan, M., & House, R. J. (2001). Cultural Acumen for the Global Manager: lessons from project GLOBE. *Organizational Dynamics, 29*(4), 289–305.

Javidan, M., House, R. J., Dorfman, P. W., Hanges, P. J., & de Luque, M. S. (2006). Conceptualizing and measuring cultures and their consequences: a comparative review of GLOBE's and Hofstede's approaches. *Journal of International Business Studies, 37*(6), 897–914.

Jones, G. K., & Davis, H. D. (2000). National culture and innovation: implications for locating global R&D operations. *Management International Review, 40*(1), 11–39.

Jones, G. K., & Teegen, H. J. (2001). Global R&D activity of U.S. MNCS: does national culture affect investment decisions? *Multinational Business Review, 9*(2), 1–7.

Kaasa, A., & Vadi, M. (2010). How does culture contribute to innovation? Evidence from European countries. *Economics of Innovation and New Technology, 19*(7), 583–604.

Kedia, B. L., Keller, R. T., & Julian, S. T. (1992). Dimensions of national culture and the productivity of R&D units. *Journal of High Technology Management, 3*(1), 1–18.

Kluger, A. N., & DeNisi, A. (1996). The effects of feedback interventions on performance: historical review, a meta-analysis and a preliminary feedback intervention theory. *Psychological Bulletin, 119*, 254–284.

Kreiser, P. M., Marino, L. D., & Weaver, K. M. (2002). Assessing the psychometric properties of the entrepreneurial orientation scale: a multi-country analysis. *Entrepreneurship: Theory and Practice, 26*(4), 71–94.

Kuhn, T. S. (1962). *The structure of scientific revolutions*. Chicago: University of Chicago Press.

Leontiev, A. A. (2006). Personality, culture, language. *Journal of Russian and East European Psychology, 44*(3), 47–56.

Li, J. (1997). Creativity in horizontal and vertical domains. *Creativity Research Journal, 10*(2-3), 107–132.

Lin, L.-H. (2009). Effects of national culture on process management and technological innovation. *Total Quality Management, 20*(12), 1287–1301.

Lubart, T. I. (1990). Creativity and cross-cultural variation. *International Journal of Psychology, 25*, 39–59.

Lubart, T. I., & Georgsdottir, A. (2004). Creativity: Developmental and cross-cultural issues. In S. Lau, A. N. N. Hui, & G. Y. C. Ng (Eds.), *Creativity: When east meets west* (pp. 23–54). Singapore: World Scientific Publishing.

Maznevski, M. L. (1994). Understanding our differences: performance in decision-making groups with diverse members. *Human Relations, 47*, 531–552.

Mead, R. (2005). *International management: Cross-cultural dimensions*. Malden, MA: Blackwell.

Meyer, A. D., & Goes, J. B. (1988). Organizational assimilation of innovations: a multilevel contextual analysis. *Academy of Management Journal, 3*(4), 897–923.

Mitchell, R. K., Smith, J. B., Morse, E. A., Seawright, K. W., Peredo, A. M., & McKenzie, B. (2002). Are entrepreneurial cognitions universal? Assessing entrepreneurial cognitions across cultures. *Entrepreneurship: Theory and Practice, 26*(4), 9–32.

Moran, R., Harris, P., & Moran, S. (2007). *Managing cultural differences: Global leadership strategies for the 21st century*. Burlington, MA: Butterworth-Heinemann.

Morris, M. H., Davis, D. L., & Allene, J. W. (1994). Fostering corporate entrepreneurship: cross-cultural comparisons of the importance of individualism versus collectivism. *Journal of International Business Studies, 25*(1), 65–89.

Mueller, S. L., & Thomas, A. S. (2001). Culture and entrepreneurial potential: a nine country study of locus of control and innovativeness. *Journal of Business Venturing, 16*, 51–75.

Nakata, C., & Sivakumar, K. (1996). National culture and new product development: an integrative review. *Journal of Marketing, 60*, 61–72.

Ng, A. K. (2001). *Why Asians are less creative than westerners*. Singapore: Prentice-Hall.

Ng, A. K., & Smith, I. (2004). Why is there a paradox in promoting creativity in the Asian classroom? In S. Lau, A. N. N. Hui, & G. Y. C. Ng (Eds.), *Creativity: When east meets west* (pp. 87–112). Singapore: World Scientific Publishing.

Noorderhaven, N., & Tidjani, B. (2001). Culture, governance, and economic performance: an explorative study with a special focus on Africa. *International Journal of Cross Cultural Management, 1*(1), 31–52.

North, D. (1990). *Institutions, institutional change and economic performance*. Cambridge: Cambridge University Press.

Peppas, S. C. (2001). Subcultural similarities and differences: an examination of US core values. *Cross Cultural Management, 8*, 59–70.

Puccio, G., & González, D. W. (2004). Nurturing creative thinking: Western approaches and eastern issues. In S. Lau, A. Hui & G. Y. C. Ng (Eds.), Creativity: When east meets west (pp. 393–428). Singapore: World Scientific.

Raina, M. K. (1999). Cross-cultural differences. In M. A. Runco & S. R. Pritzker (Eds.), *Encyclopedia of creativity* (Vol. 1, pp. 452–464). San Diego, CA: Academic Press.

Ravlin, E. C., Thomas, D. C., & Ilsev, A. (2000). Beliefs about values, status, and legitimacy in multicultural groups: influences on intragroup conflict. In P. C. Earley & H. Singh (Eds.), *Innovations in international and cross-cultural management* (pp. 17–51). Thousand Oaks, CA: Sage Publications.

Roccas, S., Sagiv, L., Schwartz, S. H., & Knafo, A. (2002). The big five personality factors and personal values. *Personality and Social Psychology Bulletin, 28*, 789–801.

Rudowicz, E. (2003). Creativity and culture: a two way interaction. *Scandinavian Journal of Educational Research, 47*(3), 273–290.

Rudowicz, E. (2004). Creativity among Chinese people: beyond Western perspective. In S. Lau, A. N. N. Hui, & G. Y. C. Ng (Eds.), *Creativity: When east meets west* (pp. 55–86). Singapore: World Scientific Publishing.

Rugman, A., & Hodgetts, R. (2003). *International business*. Harlow: Financial Times/Prentice Hall.

Runco, M. A. (2004). Personal creativity and culture. In S. Lau, A. N. N. Hui, & G. Y. C. Ng (Eds.), *Creativity: When east meets west* (pp. 9–21). Singapore: World Scientific Publishing.

Sackmann, S. A., & Phillips, M. E. (2004). Contextual influences on culture research. Shifting assumptions for new workplace realities. *International Journal of Cross Cultural Management, 4*, 370–390.

Sagiv, L., & Schwartz, S. H. (2004). Values, intelligence and client behavior in career counseling: a field study. *European Journal of Psychology of Education, 19*(3), 237–254.

Schneider, S. (1988). National vs. corporate culture: implications for human resource management. *Human Resource Management, 27*(2), 231–246.

Schneider, S., & Barsoux, J.-L. (2003a). *Management interculturel* (2nd ed.). Paris: Pearson Education Ltd.

Schneider, S. C., & Barsoux, J.-L. (2003b). *Managing across cultures*. New York: Prentice Hall/Financial Times.

Schwartz, S. H., & Sagiv, L. (1995). Identifying culture-specifics in the content and structure of values. *Journal of Cross-Cultural Psychology, 26*(1), 92–116.

Shane, S. A. (1992a). The effect of cultural differences in perceptions of transaction costs on national differences in the preference for licensing. *Management International Review, 32*(4), 295–311.

Shane, S. A. (1992b). Why do some societies invent more than others? *Journal of Business Venturing, 7*, 29–46.

Shane, S. A. (1993). Cultural influences on national rates of innovation. *Journal of Business Venturing, 8*, 59–73.

Shane, S. A. (1995). Uncertainty avoidance and the preference for innovation championing roles. *Journal of International Business Studies, 26*(1), 47–68.

Shane, S., Venkataraman, S., & MacMillan, I. (1995). Cultural differences in innovation championing strategies. *Journal of Management, 21*(5), 931–952.

Simons, G. F., Vazquez, C., & Harris, P. (1993). *Transcultural leadership: Empowering the diverse workforce*. Houston, TX: Gulf Publishing.

Smith, P. B., Peterson, M. F., & Thomas, D. C. (2008). *The handbook of cross-cultural management research*. Los Angeles: Sage.

Steensma, H. K., Marino, L., Weaver, K. M., & Dickson, P. H. (2000a). The influence of national culture on the formation of technology alliances by entrepreneurial firms. *Academy of Management Journal, 43*, 951–973.

Steensma, H. K., Marino, L., & Weaver, K. M. (2000b). Attitudes toward cooperative strategies: A cross-cultural analysis of entrepreneurs. *Journal of International Business Studies, 31*, 591–609.

Steward, J. H. (1986). Levels of socio-cultural integration: an operational concept. *Journal of Anthropological Research, 42*(3), 337–353.

Taeube, F. A. (2004). Culture, innovation, and economic development: the case of the South Indian ICT clusters. In S. Mani & H. Romijn (Eds.), *Innovation, learning, and technological dynamism of developing countries* (pp. 202–228). Tokyo: United Nations University Press.

Tayeb, M. H. (1988). *Organizations and national culture: A comparative analysis*. London: Sage.

Tellis, G. J., Prabhu, J. C., & Chandy, R. K. (2009). Radical innovation across nations: the preeminence of corporate culture. *Journal of Marketing, 73*, 3–23.

Thiederman, S. (1991). *Bridging cultural barriers for corporate success*. San Francisco: Lexington Books.

Thomas, D. C. (2008). *Cross-cultural management: Essential concepts*. Los Angeles: Sage Publications.

Tiessen, J. H. (1997). Individualism, collectivism, and entrepreneurship: a framework for international comparative research. *Journal of Business Venturing, 12*(5), 367–384.

Triandis, H. (1995). *Individualism and collectivism*. Boulder, CO: Westview Press.

Triandis, H. (1996). The psychological measurement of cultural syndromes. *American Psychologist, 51*, 407–415.

Triandis, H. (2002). Generic individualism and collectivism. In M. Gannon & K. Newman (Eds.), *The Blackwell handbook of cross-cultural management* (pp. 16–45). Oxford: Blackwell Publishers.

Trompenaars, F., & Hampden-Turner, C. (1998). *Riding the waves of culture: Understanding cultural diversity in global business* (2nd ed.). New York: McGraw Hill.

Tung, R. L. (1991). Handshakes across the sea: cross-cultural negotiating for business success. *Organizational Dynamics, 19*(3), 30–40.

Ulijn, J., & Fayolle, A. (2004). Towards cooperation between European start-ups: the position of the French, Dutch and German entrepreneurial and innovative engineer. In T. E. Brown & J. Ulijn (Eds.), *Innovation, entrepreneurship and culture: The interaction between technology, progress and economic growth* (pp. 204–232). Northampton, MA: E. Elgar Pub.

Ulijn, J., Nagel, A., & Tan, W. L. (2001). The impact of national, corporate and professional cultures on innovation: German and Dutch firms compared. *Journal of Enterprising Culture, 9*(1), 21–31.

Ulijn, J., O'Hair, D., Weggeman, M., Ledlow, G., & Hall, H. T. (2000). Innovation, corporate strategy, and cultural context: what is the mission for international business communication? *Journal of Business Communication, 37*(3), 293–316.

Usunier, J.-C. (1998). *International and cross-cultural management research.* Thousand Oaks, CA: Sage Publications.

Van Everdingen, Y. M., & Waarts, E. (2003). The effect of national culture on the adoption of innovations. *Marketing Letters, 14*, 217–232.

Weiss, A. (1984). Simple truth of Japanese manufacturing. *Harvard Business Review, 62*(July–August), 119–125.

Westwood, R. I., & Low, D. R. (2003). The multicultural muse: culture, creativity and innovation. *International Journal of Cross Cultural Management, 3*(2), 235–259.

Wycoff, J. (2003). The "Big 10" innovation killers: how to keep your innovation system alive and well. *Journal for Quality and Participation, 26*(2), 17–48.

Zander, L., & Romani, L. (2004). When nationality matters. A study of departmental, hierarchical, professional, gender and edge-based employee groupings' leadership preferences across 15 countries. *International Journal of Cross Cultural Management, 4*, 291–315.

Chapter 2
Creativity Through a Cultural Lens: The Dichotomy of "The West" and "The East"

Igor N. Dubina and Suzanna J. Ramos

Abstract Creativity is a multi-faceted phenomenon. One controversy in the creativity literature concerns whether the concept of creativity has a universal meaning or is perceived differently in various cultures. Although the fundamental idea of creativeness seems to be deeply rooted in all cultures, definitions and attributes of creativity, the level of creative activity, and the domains in which creativity is promoted, vary across cultures. In particular, the dichotomy of "the West" and "the East" is one of the most popular approaches in the characterization (at a surface level) of differences in understanding and defining creativity. However, it is not obvious what these terms exactly mean and these terms are sometimes used rather stereotypically. In this chapter, the "western" and "eastern" views on creativity are outlined as described in the literature.

Creativity and Culture: A Preamble

Creativity is a multi-faceted phenomenon. This phenomenon can be illustrated by diverse research studies in the field of creativity. One view is that creativity is an attribute of individuals (e.g. Davis 1989). Other studies include the analysis of creative production (e.g. Besemer and Treffinger 1981) as well as creativity as a cognitive process (e.g. Ward et al. 1999). Apart from the people, product and process, creativity is also understood within a social context (Mayer 1999). This suggests that the concept of creativity is inextricably linked with the social, cultural, and historical milieu.

I.N. Dubina, Ph.D., Dr.Sc. (✉)
Novosibirsk National Research State University, Novosibirsk, Russia

Altai State University, Barnaul, Russia
e-mail: igor_dubina@yahoo.com

S.J. Ramos, Ph.D.
Department of Educational Psychology, College of Education and Human Development, Texas A&M University, College Station, TX 77843, USA
e-mail: suzannaramos@gmail.com

I.N. Dubina, E.G. Carayannis (eds.), *Creativity, Innovation, and Entrepreneurship Across Cultures*, Innovation, Technology, and Knowledge Management, DOI 10.1007/978-1-4939-3261-0_2

One controversy in the creativity literature concerns whether the concept of creativity has a universal meaning or is perceived differently in various cultures. For example, some researchers believe that there is a universal understanding of the concept of creativity while another group suggests that people in different cultures perceive creativity differently.

There seems to be a major breakthrough where theories of creativity have been established based on the latter point of view. Although the fundamental idea of creativeness seems to be deeply rooted in all cultures, definitions and attributes of creativity, the level of creative activity, and the domains in which creativity is promoted, vary across cultures. Culture plays a fundamental role in defining creativity. Creativity is not only influenced by organizational factors (i.e. organizational culture), but also greatly depends on the surrounding (societal) culture as a whole (Weiner 2000).

Arieti (1976) was one of the first to suggest that potentially creative persons and creativogenic cultures are essential facets of creativity. He introduced the term "creativogenic society" to describe a type of society that enhances creativity. These socio-cultural factors are: (1) availability of cultural means (that is, an elite that has access to materials, equipment, etc.); (2) openness to cultural stimuli (cultural stimuli are not only present, but requested, desired or made available); (3) stress on becoming and not simply on being; (4) free access to cultural media; (5) freedom (or even retention of moderate discrimination after severe oppression); (6) exposure to different and contrasting cultural stimuli; (7) tolerance for, and interest in differing views; (8) interaction of significant persons; (9) promotion of incentives and awards.

From the literature, the dichotomy of "the West" and "the East" is one of the most popular approaches in the characterization (at a surface level) of differences in understanding and defining creativity. However, it is not obvious what these terms exactly mean and these terms are sometimes used rather stereotypically. The term "East" usually refers to East Asian countries like China and other countries influenced by its culture like Japan or Korea. Most published works on cross-cultural studies involve this group. Some researchers include in this group another Asian country that is not East Asian, and that is India. These mentioned countries possess general similarities in terms of the social and cultural aspects distinct from "Western" countries. One of these similarities is the tradition that traces its origin from Asian thought like Buddhism, Confucianism, Taoism, and Hinduism. The term "Western", although can be used broadly, usually refers to the US, Canada, Western Europe, Australia, and New Zealand (Weiner 2000). This conception has a long association with ancient Greece, and the ideas of Judaism, Christianity, capitalism, and rationality (Weiner 2000). In the next sections, the "western" and "eastern" views on creativity are outlined as described in the literature.

Differences in the Understanding and Definitions of Creativity and Inventiveness

In the 1960s and 1970s, the conceptual 4-P model of creativity ("Person", "Process", "Product", "Press") was suggested and developed (Rhodes 1961). However, all of these four aspects of creativity actually accent, as the most important feature of this phenomenon, the link to an observable product, which can be assessed by an appropriate group or judges, either peers or experts. For example, when creativity is considered in the "person" perspective, it is defined as the ability to produce work (*object* or *product*) that is novel and appropriate (useful, applicable, etc.). If creativity is considered in the "process" perspective, the corresponding definitions also refer to a product in the end.

The product-centered conception of creativity prevails in "the West". This conception fits with the "western" perspective on cosmic creation (Lubart and Georgsdottir 2004; Raina 1999), which involves a linear movement towards a new point and the assessment of what was created: God created *something* and God saw that what was created thing was good. Thus, the "western" conception of creativity stems from Judaic and Greek views of producing the universe by an uncreated being who brings order to the formless void (Raina 1999).

Creativity as it is understood in "the West" is rather a state of doing, not a state of being. During the last decades, creativity is considered to be a normal ability which is inherent to all normally functioning people. It is an essential life skill which includes specific cognitive, meta-cognitive and affective skills. Creative skills can be taught and enhanced through training and can be measured, in at least to some degree. Undoubtedly, creativity is considered one of the most important and principal "western" values of an enriched life.

In "the East" a product-centered creativity is less valued than a process-centered creativity (Raina 1999). The typical "eastern" conception of creativity is more focused on the process than on the result. Creativity is a personal state of being rather than an output, a connection to a primordial realm, or the expression of the inner essence of ultimate reality. Creativity is attributed less to personal factors, but rather to spiritual or social forces. Such an understanding of creativity was also proper for ancient Greece (for instance, "mania" or "enthusiasm" in Plato's sense) and, partially, for Medieval Europe, but was suspended with more individualistic conceptions during the Renaissance.

In contrast to "the West", tradition is not the antithesis of creativity. "The Eastern" creativity may take the form of intellectual or aesthetic modification, adaptation, renovation, reinterpretation, revision, reconfiguration, etc. rather than a dramatic break with tradition. For example, in "the East", artists fully respect the traditions in striving to establish their own styles and their creativity is expressed in a form of reinterpretation of the past (Leung and Leung 2004). This conception of creativity fits with the "eastern" view on cosmological process, which is characterized as an ongoing, developing or unfolding process.

The dramatically different (although some exaggerated) visions of creativity in two poles, "West" and "East", have attracted the attention of many scholars who explain them by referring to some philosophical fundamentals of "western" and "eastern" cultures. Comparing American and Japanese approaches to creativity, Herbig and Jacobs (1996) connect these differences with a historical and geographical context. The "western" view of creativity refers to the generation of new or novel ideas as a result of the competitive spirit driven by a long history of rival empires in "the West". Anything new: an idea, product or technique that provided a kingdom an advantage over a neighboring rival was rewarded and recognized. On the other hand, Japanese innovation refers to the application and refinement of an idea. By not having nearby rivals, co-operation, not competition, was espoused in Japan. Co-operation and conformity were crucial components to ensure the survival of Japanese society.

Philosophical, Religious, and Ethical Fundamentals

One of the themes in the context of cross-cultural creativity is the comparison of Socratic and Confucian philosophical traditions and intellectual and moral paradigms. Kǒng Fūzǐ (Confucius is a Latinized variant of his name) and Socrates imparted practical wisdom for their followers and founded the traditions which have often been considered as a basis to understand the differences between the "East" and "West".

"Western" thought is based on Socrates' ideas that the sole function of knowledge is self-knowledge (individualistic cognitivism), and such knowledge is the basis for a person's intellectual, ethical, and spiritual growth. Rationality, research exploration, cross-examination, public debate, and factual information are much valued (Herbig and Jacobs 1996). On the other hand, Confucius taught that the stability of society is based on unequal status relationships between people—the family is the prototype of all social organizations. A person is not primarily an individual, but rather a member of a family. Children should learn to restrain themselves, to overcome their individuality and to maintain harmony in a family and society. Values like trying to acquire skills and education, working hard, not spending more than necessary as well as patience, perseverance, and persistence closely fit this principle.

Another philosophical tradition, Zen Buddhism, is clearly imprinted in Japanese creativity and inventiveness. This philosophy does not place considerable value on rational thought. Instead, it emphasizes spiritual enlightenment and intuitive understanding. The emphasis on intuitive understanding partly explains Japan's weakness in basic scientific research where logical reasoning and systematic thinking play important roles. As a result, Japan's base of scientific knowledge and research methods has not been fully realized compared to the 'West'. This has led to entirely different meanings of enlightenment and discover in the two cultures. "*Eureka* (West) refers to the discovery of rational scientific principles while *satori* (East) means personal enlightenment" (Herbig and Jacobs 1996, p. 68).

Differences in Attributing Creativity and Inventiveness

In the context of what has been discussed so far, creativity in the "West" is associated with breaking or rejecting traditions; it is considered a discontinuous, revolutionary, relatively rapid and insightful process. Creativity is based on and involves individual traits; it is expressive, emotional, and somewhat spontaneous. Creativity is often understood as creative thinking, which should be task and method focused. Creativity is contextually pragmatic in the area of problem solving and it often tends to look outward towards "progress".

Creativity in the "East" is associated with respecting traditions and does not run contrary to them. It is considered to be a continuous, evolutionary, and slow process requiring much effort, hard work, repetition, attention, and a strong knowledge base. Creativity is based on and involves collective effort and a more structured, team-oriented approach. Creativity is understood to be socially utilitarian, since the aspects of social influence on creativity are most important—creativity should help society, improve society, and contribute to society. For example, the Chinese prefer a more practical, utilitarian conception of creativity and, as a consequence, politicians are considered the most common examples of creative individuals in China (Leung and Leung 2004).

Differences in Valuing Creativity and Inventiveness

In addition to diverse understanding of creativity, cultures value various creativity aspects differently. Creativity characteristics as defined in "the West" are less valued and encouraged in "the East" and vice versa. The principal "western" values related to creativity—individual freedom, less conformity, and self-reliance individuality, are rewarded and expected. As a result of cultivating such values, a "western" individual attempts to be open, original, and innovative. "Western" societies are more likely to produce individuals who are oriented towards newness.

The principal "Eastern" values related to creativity are social order, cooperation, duty and acceptance of an in-group authority like family, its norms and obligations; hard work and a strong knowledge base; consensus which is valued more than difference; fear of making mistakes and "losing face". As a result of cultivating such values, the tendency to delay creativity development can appear. However, the tendency of delayed creativity development and putting strong emphasis on knowledge and skills acquisition does not necessarily imply that creativity is not valued. "Eastern" societies are more likely to produce individuals who are oriented towards improvement.

Neither the cultures in the West nor the East are totally homogeneous in nature. Cultural influence can also differ between subgroups within the same culture. For example, Singapore, an Asian city, comprises three main ethnic groups—Chinese, Malays, and Indians. These groups, within an Eastern national culture, exhibit differing conceptions of what creativity entails (Ramos and Puccio 2014).

Conclusion

Concluding this article, the following are the key postulates and conclusions. No one model or approach to creativity may fit all cultures. Different cultures place emphasis on different aspects of creativity. The capability of a country to create and innovate is related to its culture. There are different culturally conditioned styles of creativity and innovation. However, the relationship between cultural values and creativity is complex as the impact of culture on creativity interplays with historical, political, and economic factors. Culture is important, but culture alone does not serve as a guarantee for a high level of creativity, inventiveness, and innovativeness. It is noted that most of the research in this field is speculative, since there are not enough significant statistical bases for such types of studies thus far. A future research area would be to systematically examine culture, creativity, and inventiveness in their dynamics by empirically investigating the relationship between changes in values and changes in rates of inventiveness.

References

Arieti, S. (1976). *Creativity: The magic synthesis*. New York: Basic Books.

Besemer, S. P., & Treffinger, D. J. (1981). Analysis of creative products: review and synthesis. *The Journal of Creative Behavior, 15*, 158–178.

Davis, J. S. (1989). Testing for creative potential. *Contemporary Educational Psychology, 14*, 257–274.

Herbig, P. & Jacobs L. (1996) Creative problem-solving styles in the USA and Japan, *International Marketing Review, 13*(2), 63–71.

Leung K., Au A., Leung, B.W.C. (2004) Creativity and Innovation: East-West Comparisions with an empasis on Chines societies, in Lau S., Hui A.N.N., and Ng, G.Y.C. (Eds.) Creativity: When East Meets West, Singapore, World Scientific Publishing Co. Pte. Ltd., pp. 113–135.

Lubart, T. I., & Georgsdottir, A. (2004). Creativity: Developmental and cross-cultural issues. In S. Lau, A. N. N. Hui, & G. Y. C. Ng (Eds.), *Creativity: When east meets west* (pp. 23–54). Singapore: World Scientific Publishing.

Mayer, R. E. (1999). Fifty years of creativity research. In R. J. Sternberg (Ed.), *Handbook of creativity* (pp. 449–460). New York: Cambridge University Press.

Raina, M. K. (1999). Cross-cultural differences. In M. A. Runco & S. R. Pritzker (Eds.), *Encyclopedia of creativity* (Vol. 1, pp. 452–464). San Diego, CA: Academic Press.

Ramos, S. J., & Puccio, G. J. (2014). Cross-cultural studies of implicit theories of creativity: a comparative analysis between the United States and the main ethnic groups in Singapore. *Creativity Research Journal, 26*(2), 223–228.

Rhodes, M. (1961). An analysis of creativity. *The Phi Delta Kappan, 42*(7), 305–310.

Ward, T. B., Smith, S. M., & Finke, R. A. (1999). Creative cognition. In R. J. Sternberg (Ed.), *Handbook of creativity* (pp. 189–212). New York: Cambridge University Press.

Weiner, R. P. (2000). *Creativity and beyond: Cultures, values, and change*. Albany, NY: State University of New York Press.

Chapter 3
Wandering the World in Search of Creativity

Robert Alan Black

Abstract Since 1977, the author of this chapter has been wandering around the world in search of creative thinking, creativity, and creative people while challenging himself to be creative everyday. Totally, he has traveled in 93 countries on six continents. This chapter shares some of the things he has discovered in his wandering.

A Preamble

Since 1977 I have been wandering around the world in search of creative thinking, creativity, and creative people while challenging myself to be creative everyday. What started out as simply a dream trip to fulfill a fantasy of a lifetime, travel around the world like Phileas T. Fogg did in Around the World in Eighty Days, became partially a research project.

Initially, I was searching architecture, graphic design, signage, and interior design. In 1998 I began searching for examples of creativity being taught and trained, and in 2001 I took my first of ten trips completely around the world in search of creative thinking, creativity, and creative people while challenging myself to be creative everyday. I have traveled in 93 countries on six continents. This chapter (that based on my initial attempts at preparing this article for the 2002 Creativity's Global Correspondents) shares some of the things I have discovered in my wandering.

Since 1976 I have focused much to most of my efforts (personally and professionally) on the application and development of creativity: mine and others. Over the past 39 years, professionally, I have slowly refined my focus to S.P.R.E.A.D.ng™ creative thinking throughout workplaces. S.P.R.E.A.D.ng™ is the acronym I use to demonstrate to people what I believe needs to be done within workplaces to enhance, expand, and enrich the creative thinking of all employees (support, promote, recognize, encourage, apply, and develop).

R.A. Black, Ph.D. (✉)
S.P.R.E.A.D.ng Creative Thinking Throughout, Athens, GA 30604, USA
e-mail: alaniscre8ng@gmail.com

I.N. Dubina, E.G. Carayannis (eds.), *Creativity, Innovation, and Entrepreneurship Across Cultures*, Innovation, Technology, and Knowledge Management, DOI 10.1007/978-1-4939-3261-0_3

During my 73 days traveling around the world, the summer of 2001, through New Zealand, Australia, Malaysia, Singapore, Sri Lanka, India, Dubai, Turkey, Denmark, Germany, the Netherlands, England, and France, I sought to find examples of creativity everywhere I went and to interview people about their own creativity and creativity in general in their country.

As a skeleton structure for this chapter, I have chosen to use two lists of traits of creative people. The first list comes from the TTCT™ (Torrance Tests of Creative Thinking), and the second comes from an ongoing project I have been working on since I was a doctoral student studying with Paul Torrance in 1980. The 20 traits from the TTCT™ represent 20 traits that are examined by the TTCT™ tests based on over 40 years of E. Paul Torrance's scientific study with subjects around the world. The 32 traits from my "Are You a Crayon Breaker?" exercise come from a survey study I did of articles on the traits of creative people written from 1950 to 1980. I have used the survey as part of over 2100 professional speeches and workshops since 1981 to suggest the existence of creative thinking potential in all people and also to indicate different styles of creative thinking.

During my trip, I often had people I met and interviewed complete the 32 traits survey to share which they believed fit them. While we talked together, I introduced them to the work and ideas of E. Paul Torrance and the 20 traits from the TTCT™. In addition, as a post trip survey, I have reviewed my daily journal notes using both the list of 20 from the TTCT™ and my 32 "Crayon Breaker" exercise traits for potential understandings of what I experienced and discovered.

The Traits

TTCT™ Traits

1. Fluency—many ideas
2. Flexibility—different types of ideas
3. Elaboration—addition of details
4. Originality—uniqueness
5. Abstractness of approach moving from reality
6. Openness—resisting early closure or completion
7. Change of context (cross-interpretation)
8. Combination of ideas/facts—synthesis
9. Breakthrough from current limits
10. Unusual viewpoint
11. Internal perspective
12. Humorous perspective
13. Richness and colorful detail

14. Feelings and emotions
15. Fantasy
16. Movement and sound—sense change
17. Multiple idea combinations
18. Macroscale perspective—seeing from larger view
19. Provocative viewpoint
20. Future orientation

Are You a Crayon Breaker™

1. Sensitive
2. Not motivated by money
3. Sense of destiny
4. Adaptable
5. Tolerant of ambiguity
6. Observant
7. Perceive world differently
8. See possibilities
9. Question asker
10. Can synthesize correctly often intuitively
11. Able to fantasize
12. Flexible
13. Fluent
14. Imaginative
15. Intuitive
16. Original
17. Ingenious
18. Energetic
19. Sense of humor
20. Self-actualizing
21. Self-disciplined
22. Self-knowledgeable
23. Specific interests
24. Divergent thinker
25. Curious
26. Open-ended
27. Independent
28. Severely critical
29. Nonconforming
30. Confident
31. Risk taker
32. Persistent

The Beginning of My Creative Wanderings

From 3:30 pm in Athens, Georgia on June 25th 2001 to 5:30 am in Auckland and finally 11:00 am June 27th in Christchurch I began my first trip around the world.

After I arrived in New Zealand, in Christchurch, on the train to Dunedin, in Wellington, and in Auckland, I asked architects, designers, advertising art directors and account executives, landscape architects, and theater people I met and/or stayed with to complete my "Crayon Breaker" survey.

Initially, I discovered that the lowest number of traits checked off by the people was 15 with a few marking all 32, including #28, severely critical. Over the 35+ years I have been using the exercise, the people in my programs or audiences have generally marked between 5 and 15 with a few, most times, who mark over 15.

Most people do not openly admit to this. It is the one trait I discover in personal interviews and reviews of biographies and autobiographies of highly creative people, living and dead. My assessment is that the "higher" creative people are severely critical of three things and are accused of being severely critical of a fourth. They tend to be severely critical of (1) themselves, (2) their work, and (3) the potential of their fields of passion. Because of these three, people who are much less creative as them see the "higher" creatives as being severely critical of other people. My experiences and ongoing study do not support that anywhere in the 93 countries I have traveled in during my life.[1]

During my interviews in New Zealand after the 32 traits survey was completed and reviewed, I discussed the TTCT™ traits as trainable/learnable traits asking the people who were company owners or managers, if they actually, consciously, strived to increase the creative thinking abilities and skills of their people. Except in a few isolated cases where the interviewees were creative thinking consultants, the answer was always no, and generally the people were unaware that creative thinking could be increased or taught.

My 4-week journey around Australia (six states and two territories) took me to Sydney, Canberra, small towns and a self-sufficiency site on a mountain in the Snowy Mountains, Melbourne, Hobart, Adelaide, Uluru, Alice Springs, Perth, Darwin and a couple small towns in the Northern Territory, Cairns, and finally Brisbane.

In most of the cities and towns, I stayed with Servas members, an international travel organization whose members open their homes to other members as their guests. Using the Servas Australia directory as I had in New Zealand, I hand selected a mix of creative people to stay within each city with no repeats of professions or occupations. They consisted of creative thinking consultants, designers, theater promoters/directors, therapists, ceramic artists, counselors, writers, fabric artists, sculptors, and teachers or trainers.

---

[1] At the moment of the 2001 travel (*Ed.*)

Researcher Becomes Searcher (Part I)

As my trip continued, I became less systematic with my data collection primarily because I became much more involved in getting to know my hosts and the people I met along the way instead of playing scientific researcher, and also I become more involved in simply the various experiences day by day.

Periodically in Australia, I had people complete the "Crayon Breaker" survey. The results were the same: very high numbers of traits selected. I continued my sharing information about E. Paul Torrance's work and some about my own and other creative thinking consultants I have gotten to know from various countries and ones I got to meet along the way during the journey.

A Primary Creative Learning: Dealing with Daily Frustrations

One learning that kept coming back again and again throughout my trip was one I learned from Joel Goodman from the Humor Project many years ago: "If when something happens you can say 'some day I'll laugh about this' then why not start now!" When frustrating and highly stress producing or simply very negative things happened during my travels around Australia and then in other countries later, I would "step out of my shoes, boots, or saddles" and remind myself of Joel's bit of wisdom, and within a few moments I was smiling and laughing and making notes of how to turn the experience or situation into material for a future article or speech.

Learning from the Weather

Have you ever thought that it takes more creativeness to enjoy a rainy day than a sunny one?

Instead of becoming frustrated by the many days of rain I experienced in the southern states and territories of Australia, I chose to use my creativeness to turn them into wandering adventures.

One example would be my third day in Sydney. It had rained off and on, mostly on, for the entire 3 days making it difficult to capture the beautiful sites and experiences with my point and shoot Fuji camera. That day after the first couple of hours of riding one harbor boat after another, there are a series of boat lines that crisscross Sydney Harbour from one end to the other, I noticed that the streets were becoming extremely crowded with people. All the while it began raining harder and harder.

Most of my life I experienced claustrophobic-like reactions in crowds of people, especially when the people all seem to have gotten up that morning with the sole intention of getting in my way. That day by noon, it had gotten worse and worse. I felt like a young chicken stuffed into an extremely wet and overly crowded chicken growing house with chickens all around and over me, so packed in a can of sardines would seem vastly loose.

I pushed my way through the crowd and found a train station under one of the high-rise blocks of buildings. On the spot I had decided to go out to the 2000 Olympic Site just to get away from the mobs of people. As I pushed my way through, I asked person after person how to get to the Olympic Site. After seven or eight different answers, I just got on a train heading out of town and asked the conductor once I sat down. My luck was with me. The train I had run onto was the right one.

In about 20 min, I was walking in the vast, very, very open area of the Olympic Site enjoying the environment and architecture. One person per 20 acres instead of the thousands per 200 square feet I had just left behind in Sydney. After enjoying the openness and viewing many of the contemporary sports arenas, I realized that the only building, way off in the distance, almost to the horizon, that was open had a very long line wrapping nearly around it. So I decided to head back to Sydney to catch the bus from downtown Sydney, near the Opera House, to my creative thinking consultant friends' home in the northern suburbs and end my day of wandering in Sydney.

I slowly walked back to the empty train station. There was a good reason it was empty. The last train back to Sydney that evening was leaving in 3 min.

Learning: Trust Your Subconscious

Once again my subconscious or intuition had taken control and I had unknowingly trusted it to guide me.

That was another learning/relearning I discovered as I traveled throughout the journey: trust my subconscious and intuitive skills.

Learning: Trusting Some Natural Creative Traits

In Hobart, Tasmania, while staying with Helen and Andre, two very successful and accomplished ceramic artists, I experienced the value of trusting the natural creative traits of curiosity, exploration, divergence, openness to premature closure, independence, imagination, and others. Because I was trying to experience something of each of the eight states and territories all within 4 weeks, I had only planned to be in Hobart and Tasmania 3 days and 2 nights.

Tasmania is an absolutely beautiful island state, which until this summer I mistakenly had thought it was a separate nation. By only planning to be there for such a short time, I was not going to see or experience much of the natural beauty of the island on the ground: the vast forests, valleys, rivers, mountains, snowy peaks, etc. Add to that it was raining most of the time as well.

So I played tourist and gathered up maps and went to the Chamber of Commerce plus the Tasmanian Tourist Agency offices to pick up information of what I might see in less than 48 h by bus (commercial or tour), bicycle, or foot.

After a couple hours of frustration of trying to make my time work, I simply decided to enjoy Hobart by foot and plan on returning to Tasmania in the future for a much longer time. I was scheduled to speak in the afternoon on adding creative thinking to your life and I wasn't using my own.

I threw away all but a simple street map of the downtown area of Hobart and began to "wander." It became another "relearning"—allow yourself to creatively experience life instead of always trying to create it. That day and the next 1/2 day became fantastic. I experienced many people, the streets of Hobart, the interiors of many shops, restaurants, and much urban art and toured several artist studios enjoying a great variety of art.

"Letting go and experience the creativeness and creativity that surrounds you" became a creative tool throughout the remainder of my 73-wandering journey because of my time in Hobart.

Learning: Wandering Without a Predetermined Plan

A learning that I often share with participants and students is that of simply "wandering" and letting the creativity that surrounds them remove the "clouds" or blocks of creativity that prevent them from being creative at any given moment. This I did in shops, malls, streets, banks watching the customers, waiting around ATM machines, ceramic, painting, fabric, sculpture studios, toy stores, grocery stores, along piers, and restaurants.

Wandering: My Greatest Creative Tool

After Hobart, it was Adelaide, Uluru (Ayers Rock, Alice Springs, Indian Pacific) train across from Adelaide to Perth, flights to Darwin, Cairns, and Brisbane. The experiences and lessons during those 3 weeks continued to reinforce what I had already lived.

Learning: Experiencing Varied and Many Cultures to Expand Creativeness

By the time I reached Brisbane, my last destination in Australia, located in Queensland, I had already experienced many cultures and subcultures. That was one of my earliest creative learnings from 1977 when I first took an extended trip involving visiting 20+ countries in Europe, Eastern Europe, and North Africa.

Learning: To Expand and Enrich Our Creativeness and Creative Thinking Skills, We Need Only Expose Ourselves to Varied Cultures and Peoples

From Athens, Georgia, on June 25 until August 8 in Brisbane, I had experienced New Zealanders, natives and immigrants, from the very northern part of the North Island to the southern section of the South Island, Maori natives who live in a variety of ways from very old custom to very contemporary. I had traveled thousands of miles by train, bus, cab, foot, and plane experiencing Australians from all eight states and territories: New South Wales, Victoria, Canberra, Tasmania, South Australia, Western Australia, the Northern Territory, and Queensland. They ranged from very contemporary residents of Sydney, Melbourne, Adelaide, Perth, and Brisbane. Throw in hedonistis who enjoy their lives in Darwin, Cairns, or along the Gold Coast between Cairns and Brisbane, some making their livings as scuba diving instructors or street artists. Also they included self-sufficiency living people from the Snowy Mountains who work only when the money runs out plus country people from each of the eight areas.

Add to that were outbackers who spend days and weeks totally alone in the barren outback. Plus include a mix of Aborigines from those who live as their forefathers and mothers have lived from 40,000 years to college-educated professionals who had been trained by both their native cultures and the white culture of modern Australia.

Varied Traits Provoked Through Cultural Immersion

The learnings from such immersion in varied cultures help to expand, enrich, and provoke increased creativeness through the following traits:

- Abstractness of approach moving from reality—seen through the art and thinking of so many different peoples
- Adaptable—experiencing how so many different Australians live their lives
- Breakthrough from current limits—caused by the contrast of my culture with so many others
- Change of context (cross-interpretation)—continually exposing myself to daily to by the hour changes of context and culture
- Combination of ideas/facts (synthesis)—trying to create a synthesis of everything I was learning and experiencing
- Synthesize correctly often intuitively—trusting these abilities in myself by the hour and day
- Curiosity—pushing this to extreme limits everywhere I went whether flying by helicopter to the top of a glacier in Franz Josef, New Zealand, or a seaplane over the skyline of Darwin, Northern Territory, Australia, or walking around the famous Uluru Rock

- Divergent thinking—being open to experiencing this everywhere I went
- Open-endedness—reminding myself not to go to premature closure quickly
- Elaboration—learning to see through the eyes of others
- Fantasy—trying to experience the fantasy lives of other cultures
- Multiple orientations—past, present, future, and virtual
- Internal perspective—trying to experience these in others and myself
- Question asker—being willing to ask and be asked
- Richness and colorful detail—experience vast varieties everywhere I went
- Risk taker—opening myself to risk taking daily or by the hour
- See possibilities—opening myself to possibilities everywhere
- Unusual and provocative viewpoints—opening myself to these everywhere

Researcher Becomes Searcher (Part II)

Changing Cultures

During the first 6 weeks, I traveled where English, at least some version of English, was spoken wherever I was. When I left Brisbane for Kuala Lumpur, I left that security blanket behind me at least part of the time each day.

The Physical Environment and Its Creativity

Kuala Lumpur and Singapore have much in common. Both have worked very hard to enter the twentieth and twenty-first centuries in less than 40 years each. Both skylines are filled with the most contemporary buildings any architect could dream of. Each of the major cities I had been in so far in New Zealand and Australia also was filled with many relatively new buildings, with Brisbane having the largest concentration, mostly built in the 1990s and Sydney slightly behind it because of it hosting the 2000 Olympics.

What Kuala Lumpur and Singapore still possess that neither New Zealand nor Australia do not is also the ancient. NZ and Australia are barely 200 years old. Both KL and Singapore are also ancient countries filled with temples, shrines, and slums, by today's standards, that use construction types and living styles that people have lived for over 3000 years.

Learnings: Western Creative Meets Eastern Creativity

The learnings for me as an outsider with little to no knowledge of the many cultures that have lived and died in Malaysia and Singapore were that of watching the contrasts which produce the richness along with the vast confusion and stress that appear to exist in both of these cultures.

Seeing the simple lines, colors, and forms of the newest contemporary buildings and developments contrasted against the extremely complex, ornately detailed, and polycolored Hindu temples caused me to recognize the need for juxtapositioning of our thinking.

Onto Sri Lanka to Learn More: Creativity During Revolution

From Singapore after about 10 days spent traveling back and forth between KL and Singapore off I went to rebel-torn Sri Lanka, the paradise that has drawn many people from around the world. Less than 2 weeks before I arrived, rebel forces blew up five commercial airplanes on the runways at the airport. I was traveling to Colombo, Sri Lanka, with the purpose of presenting professional programs on creative thinking in their workplaces and touring a little.

From my arrival at 12:30 am to be picked up by a total stranger to be driven through totally dark streets to a hotel that I only knew was located somewhere in Colombo, the capital city where I would sleep and finally meet my formally unmet client in the morning, I needed to use my creative skills to learn to accept and let go of my growing fears.

Learning: Our Cultures as Blocks to Our Creativity

The greatest learning for me among many from my 4 days and 3 nights in Sri Lanka was the power of blocks upon the creativity of a complete culture of people. With all the roadblocks, checkpoints, military personnel, nightly curfews, and daily required power outages, I saw creativity everywhere I went from how to drive effectively in a non-geometric fashion to get from point a to point b through absolute chaos without traffic lights or electric auto turn signals to some of the greatest creative lunch and dinner buffets I have ever experienced. True I felt like a dragon with a flaming mouth most of the time I was eating, but I did learn how to enjoy even the pain of spice as long as I had a glass of fresh orange juice, something sweet, or ice cream to contrast the spices in my mouth with.

When I reacted to my driver's actions by putting up my hands to cover my eyes from the possible car crashes at every turn, I was laughingly warned to watch out for the drivers in India. The warning was truly well given.

Going Deeper into the East Like Marco Polo in Reverse

From Colombo, I went on to Chennai, India (once Madras), located near the southeastern tip of the Indian peninsula. Chennai is a very ancient city that too tries to mix the ancient and the near contemporary. Ancient construction techniques are used to build the local EDS office building or the local Domino's Pizza delivery shop.

Learning from the East

Visiting countries such as Sri Lanka and India as a citizen of the USA is a learning in itself. From our nearly anal obsession with geometrically laid out streets and driving laws to the totally amorphic conditions on their streets and apparently nonexistent driving laws, you truly experience the contrast between focused convergence and seemingly aimless divergence.

Researcher Becomes Searcher (Part III)

Going Home: Again to Istanbul

From Chennai I traveled to Istanbul, Turkey, stopping for a few hours in Dubai, the most modern convention/conference mecca of the world. In the 140° desert of Dubai lies the most modern airport with the largest and most contemporary duty-free mall, not shop or shops, a full-range mall. One lesson from those 2 h was that creativity can occur in any environment no matter how harsh or repressive to human existence.

Arriving in Istanbul was a pleasure, partly because I was being picked up by a friend I would spend much of my time with in her beautiful city and because I had been there three times before. I was not going to experience the shocks of the unknown or simply imagined environments and cultures of Malaysia, Singapore, Sri Lanka, and India.

Instead, I was returning to a favorite city that I have traveled extensively about with friends and alone.

Learning: Integration of Ancient, Old, Current, and Future

One of the extensive creative learnings that Istanbul provided was the integration of ancient, old, current, and future cultures, religions, and peoples. The most distinct difference at first for westerners is that the population is over 90 % Muslim with a very small minority of Christians or Jews. Yet there is a sense of peace that exists in the initially appearing chaos.

Wandering to Replenish My Creative Soul

This time in Istanbul I chose to walk or boat most places when I wasn't riding in my friend's car. Wandering through the streets basically unnoticed was a pleasure. It enabled me to explore, experience, and examine the ongoing creativity that surrounded me in the Bazaar, the commercial areas where the wholesale trading goes on, or some of the most contemporary and pricey malls I have ever seen in the world. Combined with that were several strolls along the Bosphorus Straits on both sides, European and Asian.

Learning: Integrating Contrasts

A significant learning from Istanbul comes from the fact that it is a city of over 20–24 million people and is the only city that exists in two separate continents connected by bridges. The learning is the value of contrasts and integration of contrasts to produce creativity and innovative ideas and solutions.

Revolving Back to Western World

Time to contrast again, a flight from Istanbul to Copenhagen. From ancient chaos to modern and contemporary highly controlled order. From high contrasts in nearly every aspect of life to sameness and uniformity, much by law and culture.

Learning: Control and Orderliness Can Produce Creativity

My first learning from returning to Copenhagen after 24 years was how controlled and orderly it is and apparently lacking in spontaneity and creativity. It took about a day to clear up my creative blinders from Istanbul, Chennai, Colombo, and Kuala Lumpur and to begin to see the wonderful creativity in Copenhagen and the countryside of Denmark.

The learning, no matter how much control and systemization a culture may generate the natural desire for creativity, will show through. In Copenhagen, individual creativity does not seem to exist initially until you begin to look for it and become more open to experiencing it. Doors, doorways, entrances, window flower boxes, personal window displays, gardens, graphics, furniture, silverware, artwork, and ceramics from artistic to everyday chinaware are what demonstrates the creativity of the Danish people as individuals and not just members of a highly refined creative nation.

After once again seeing their creativity, I felt more relaxed. Then I went with an Italian friend, who had arranged to be in Copenhagen while I was there, out to dinner to walk the streets at night and to visit Tivoli, one of the oldest amusement parks in the world. Viola, the internal creativity and desire for independence and divergency showed inside Tivoli even until 2 or 3 in the morning.

From Refined Control to Refined Individualism with Control

From Copenhagen by train, boat, bus, foot, and car, I traveled for an entire day to Delft in the Netherlands across northern Germany to be welcomed by my cyberspace friend, Marc Tassoul, professor of creativity and industrial design plus a consultant and creator of the first creativity-focused Internet email discussion list, CREA-CPS.

Prior to this trip, I had the pleasure of visiting and staying in Delft several times beginning in 1977. Delft is a walking town, medieval architecture and design combined with the most contemporary available in the world. Each integrated beautifully at a human scale. No high-rise buildings within the city. Everything is located a few minutes away by foot.

Learning: Vary the Scale and Speed to Rejuvenate Creativity

The learning for me was in the need to vary the scale and speed of our lives to help expose, expand, and enrich our natural and developable creative thinking traits and skills.

Delft is an example of a completely designed and integrated community. All that is built new is integrated to create a harmony that is rarely experienced in any other community, town, city, or country.

Off to the Center of the British Empire: Shakespeare's Home

After an abundance of sensory enjoyment and ample time with friends, I was off to England, Stratford specifically, to experience another small-scale well designed community. What Stratford lacks in contemporary design it relishes in historic and singularly focused creatively. It is an entire community devoted to the creativity of one man, William Shakespeare. Yet within and among all the curios and tourist attractions is the love of the beauty of the landscape and the blending of the architecture and contemporary life.

Learning: Change of Scenery Can Relish and Replenish Creative Spirit

A learning for me from Stratford is that I can enjoy intensely crowded streets and parks during the mid afternoon while being able to jump on my rented bicycle to travel off to the countryside in a very few minutes getting lost in the beauty of a travel along the Avon that William and Anne and their children probably walked hundreds of years ago. I can also bicycle or walk the same busy streets at dawn or late at night after the evening's Shakespeare Theater performance as if I was the only person on earth.

Learning: Juxtapositioning Imagination

Added to that learning is the ease with which I can play with my imagination in a town like Stratford-on-Avon juxtapositioning my thoughts from contemporary life, talking with a political cartoonist I met on the train ride to Stratford to pretending I am a citizen of medieval time Stratford walking to experiencing William's latest creation at the theater.

Juxtapositioning in Time

From Stratford-on-Avon, I traveled by train to Salisbury to meet a met cyberspace creativity friend, John Thomas, a retired teacher and creativity author, and to return again for the third time to the time of the Druids at Stonehenge.

Every Town Possesses Vast Contrasts of Creativity

Salisbury provided a sampling for contrasting time comparing the famous Salisbury Cathedral to the Druid Circle of Stones at Stonehenge to experience distinctly different spiritual expressions of creativity. Walking the now controlled, physically and electronically, route around the world famous stones, then walking slowly around the famous cathedral both at midnight under the spotlights and in the early morning at daybreak simply experiencing both provided me a "creative soul" fill-up.

Back to Modern Times

Off to London next, not really wanting to be there, except to meet up with some highly creative people at a creative consultant firm and St. Luke's advertising agency, reported to be one of the most creative in the world today was my next planned destination.

Letting Go Once Again to Learn

Thank you goes to Joel Goodman again. I chose to laugh and to enjoy the on and off rain and chose to walk aimlessly the first day before meeting my contact at What If?!. From my visit at the offices of What If?!, I received several tips for what to see in London today. I combined that with on and off rides around London via a constantly available series of double-decker bus tour around the entire city. Instead of being frustrated by the scale and density of London, I fell in love with experiencing parks from small squares to Hyde Park in size, art galleries with the work of Picasso, Dali, and many yet to be known as famous artists, the British Air Eye gigantic Ferris wheel, etc.

Learning: Changing Perspectives Without Specific Plan

A learning from London this time came in the enjoyment and creative potential of constantly changing perspectives and scale, both deliberately and spontaneously as the spirit moved me, from touring the reconstructed Globe Theatre to walking along both sides of the Thames to the Tate Gallery to a seafood festival to a street musician playing an ancient Chinese instrument to a mix of varied food cultures.

What started out as depressing 3 days turned into fantastic and richly creative days.

It Can't Be Done!: Tunneling to Even More Creativity

Then it was off through one of the greatest examples of engineering creativity in Europe, the England to France tunnel under the English Channel. The learning: what seems impossible with today's abilities can become easy with tomorrow's.

Arriving in Paris is always fun. This was my seventh visit to Paris. My goal was to complete my trip in a beautiful city and give a speech to the newly formed French Speakers Association in Paris on my last night at the American Church along the Seine on creative thinking as a professional speaker.

Learning: Complete Openness to Experience Breeds Creativity

My 4 days and 3 nights in Paris and the surrounding area including many of its suburbs, Chartres, and Disneyland Paris were nearly completely spontaneously filled with creative adventures and complete openness to experiences as they happened, minute by minute, hour by hour, person by person, and experience by experience. Included was walking from the front door of my hotel with a complete view of the Eiffel Tower only a few blocks away, a view I would experience many times throughout the time I was there day and night to returning each evening to sleep once again.

The overall learning from my time in Paris this time was to remind myself to set basic goals, targets, that fulfill my mission and fit my vision while being open to experiencing life as fully as possible at all times during the day each day.

Returning Home Once Again

73 days in search of creativity, creative thinking, and creative people from Athens, Georgia to Auckland to Paris, and back to Athens again, I found and experienced each of the three everywhere I went and so will all of us if we simply apply the natural traits of creative people and creative learnings that we choose to develop throughout our lives.

Chapter 4
The Geography of the Creative Mind: A Cross-Cultural Study of Implicit Theories of Creativity Between the USA and Singapore

Suzanna J. Ramos

Abstract The "term" culture does not have a unilateral definition. Culture allows us to define who we are and what is meaningful, as well as to manage our physical and social milieu. As a result, our cultures have a tremendous influence on the way we think and feel, the way we view the world, the way we communicate, and the way we behave. Culture is not a static construct but created daily through contacts, exchanges, and communication between individuals and their social landscapes. The underlying assumption is that people's thoughts and actions are guided by their own personal definitions of creativity, and they have their own beliefs about how to foster and judge creativity, which may be similar to the theories developed by experts in the field of creativity. This chapter explores the extent of influence of culture on implicit theories of creativity among laypeople from the USA, a predominantly Western culture, and Singapore, a predominantly Eastern culture, as well as the ethnic groups in Singapore, in regard to adaptive and innovative styles of creativity and their own conceptions of creativity.

Introduction

Early research on creativity tended to adopt an individualistic perspective, where creativity has been theorized in a variety of domains. Some examples include creativity as a process that occurred in the minds of individuals who possessed suitable personal characteristics and experiences (MacKinnon 1965), cognitive approaches in terms of cognitive style (Martinsen and Kaufmann 1999), and the pragmatic approach, where the concern is primarily with developing creativity (De Bono 1971; Osborn 1953). Glaveanu (2014) refers to this perspective as the "I-Paradigm," where the individual is the unit of analysis.

S.J. Ramos, Ph.D. (✉)
Department of Educational Psychology, College of Education and Human Development, Texas A&M University, College Station, TX 77843, USA
e-mail: suzannaramos@gmail.com

I.N. Dubina, E.G. Carayannis (eds.), *Creativity, Innovation, and Entrepreneurship Across Cultures*, Innovation, Technology, and Knowledge Management, DOI 10.1007/978-1-4939-3261-0_4

The "voice" behind the "I-Paradigm" was that of J.P. Guilford, in his 1950 American Psychological Association's presidential address. He made a clarion call to psychologists to make creativity a focal point of psychological inquiry (Guilford 1950). Following Guilford's message, psychologists responded to this call and creativity research flourished in the 1960s and early 1970s. The literature on creativity included several core disciplines of psychology, mainly personal attributes, cognitive processes, and the acquisition and actualization of creative potential (Simonton 2000).

It was only in the 1980s and 1990s that interest in the role of culture in creativity studies gained momentum, since "creative expression is a universal human phenomenon that is firmly grounded in culture and has its own profound effect on culture itself" (Rudowicz 2003, p. 273). Sociologists and anthropologists have long pointed out that creativity is mostly a sociocultural phenomenon (e.g., Kroeber 1944). Furthermore, Csikszentmihalyi (1990) asserts that to study creativity alone is "like trying to understand how an apple tree produces fruit by looking only at the tree and ignoring the sun and soil that supports its life" (p. 203). In sum, in studying creativity, one must consider the holistic nature of the individual as part of an evolving system within a cultural setting.

The "term" culture does not have a unilateral definition. Stated simply, culture is a "set of attitudes, values, beliefs, and behaviors, shared by a group of people, communicated from one generation to the next via language or some other means of communication" (Matsumoto 1994, p. 4). Culture allows us to define who we are and what is meaningful, as well as to manage our physical and social milieu. As a result, our cultures have a tremendous influence on the way we think and feel, the way we view the world, the way we communicate, and the way we behave. Culture is not a static construct but created daily through contacts, exchanges, and communication between individuals and their social landscapes.

Apart from using psychometric approaches to measure creativity, a useful application of a person-oriented psychometric method is the role of implicit theories. Unlike explicit theories where they are "opinions and views held by scientists" (Runco 1999, p. 27) and typically based on "some psychological or scientific construct" (Runco 1990, p. 236), implicit theories are tacit knowledge held by an individual and are often "personal rather than shared" (Runco 1999, p. 27). Davis (2004) describes implicit theories as conceptions held in people's minds and can serve as "mental prototypes that can be used to decide if a product, behavior or person is creative" (p. 70). Therefore, the underlying assumption is that people's thoughts and actions are guided by their own personal definitions of creativity, and they have their own beliefs about how to foster and judge creativity, which may be similar to the theories developed by experts in the field of creativity.

In order to discover laypeople's implicit theories of creativity that can lead to greater insights on a present explicit theory of creativity, this study utilized the explicit theory of Kirton's (1976) adaption and innovation theory (KAI). Kirton's (1976) explicit theoretical proposition is that individuals lie within a cognitive style continuum ranging from adaptive to innovative orientation. At one end of the continuum is the high adaptor, who tends to accept the problem and stay within the current paradigms, rules, policies, and structures. They work to improve on them and generate solutions that are conventional, less disruptive, and easier to implement. At the other end of the continuum is the high innovator, who tends to abandon the current paradigm and redefines the problem with a new approach. Thus, adaptors do things better, while innovators do things differently when solving problems (Kirton 1976, 1999). The assumption of this study is that if there is indeed a matching between these two types of theories (the implicit theories of laypeople and the explicit theory of the KAI), laypeople will have an innate understanding that they are creative but in different ways within the continuum of an adaptor or innovator.

This chapter explores the extent of influence of culture on implicit theories of creativity among laypeople from the USA, a predominantly Western culture, and Singapore, a predominantly Eastern culture, as well as the ethnic groups in Singapore, in regard to adaptive and innovative styles of creativity and their own conceptions of creativity. Although there have been comparative studies between a Western culture and an Eastern culture (Kim 2005; Li 1997; Soh 1999), there have not been studies done in regard to ethnic groups within a particular national culture. In this study, Singapore, as a national culture, is multiracial in nature because it comprises three main ethnic groups—the Chinese, the Malays, and the Indians. Comparisons among these ethnic groups provide deeper insight as to whether issues of ethnicity and other cultural mores distinct in each ethnic group play a role in how creativity is conceived.

The following were the specific research questions that guided this study:

1. Using Kirton's explicit theory of adaption and innovation (KAI) to access laypeople's implicit views of creativity
    - To what extent do laypeople from the USA and Singapore have similar views of Kirton's contention that adaptors and innovators are equally creative?
    - To what extent do different ethnic groups within Singapore (i.e., Chinese, Malays, and Indians) have similar views of Kirton's contention that adaptors and innovators are equally creative?
2. When asked to define creativity in their own words
    - To what extent do laypeople from different national cultures in the USA and Singapore hold similar or different conceptions of creativity?
    - To what extent do laypeople from different ethnic groups in Singapore hold similar or different conceptions of creativity?

Method

Participants

The study included 523 participants, ranging from 18 to 75 years of age. The participants were obtained through convenience sampling. There were three sets of samples. The first set was sample A, which consisted of 139 participants from the USA, in Buffalo, New York. The second set was sample B, which consisted of 199 participants from Singapore. As for the third set, sample C, it consisted of 185 Singaporean participants from the three main ethnic groups—84 Chinese, 54 Malays, and 47 Indians. All the participants had no formal training or background in creativity studies, as well as no prior knowledge of Kirton's adaption and innovation (KAI) theory.

Materials

The study utilized a questionnaire that contained a close-ended section and an open-ended section. In the close-ended section, the participants were given descriptions of two different people. The two descriptions were characteristics of the adaptor and innovator, which were directly taken from Kirton's work (1994). The two sets of characteristics were labeled as person A and person B. Approximately half the questionnaires had characteristics of the innovator and labeled as person A, while the other half had characteristics of the innovator but labeled as person B. This arrangement helped to suppress any bias and counterbalance the effect of reading one description first and, for that reason, rating one person higher than the other. A response scale ranged from 1 to 10 (one meaning "not at all creative" and ten meaning "exceptionally creative").

The questionnaire also included an open-ended question. The question was: "When you hear the word 'creativity,' what words come into your mind? Please list below those words you associate with creativity." Overall, the survey was designed to take less than 10 minutes to complete.

Procedure

The researcher worked remotely from the USA with a research assistant based in Singapore and another research assistant based in Buffalo, New York. The researcher conducted online discussions with the two research assistants so that they were familiar with the goals and procedures of the study. Ethical considerations like voluntary participation and use of consent forms were thoroughly discussed and explained. They were given a detailed description of the study on paper so that they were familiar with the procedures for obtaining participants for the study.

Before engaging an individual to participate in the study, the conditions of participation in the study were explained. The participants read and signed the consent form and indicated their gender, occupation, and age on the front cover of the questionnaire. Participants from the Singaporean sample (sample C) also indicated their ethnicity—Chinese, Malay, or Indian. Care was taken to ensure that the Singaporean sample consisted of only Singapore citizens, as the country has a large proportion of permanent residents from various countries. The forms were in English as this was the lingua franca, so translation to the various languages was not necessary.

Once the participants filled out the consent form, they proceeded to the questionnaire. They were given as much time as they needed to complete the full questionnaire. Overall, the questionnaire took less than 10 minutes to complete.

Results

The first analysis of the close-ended questions in the questionnaire used *t*-tests to compare the participants' ratings of the adaptor and innovator across the three samples. Table 4.1 indicates the minimum and maximum ratings of the adaptor and innovator styles, the mean adaptor and innovator ratings, and the standard deviations of each sample, A, B, and C.

Table 4.1 shows the mean rating for the adaptive style ranged from 4.6 to 4.9, while the mean rating for the innovative style ranged from 7.1 to 7.3. In all three samples, the innovator style received higher ratings for creativity. It is also noted that both adaptor and innovator styles received ratings across the full continuum; that is, both the adaptor and innovator styles were rated as 1 (not at all creative) and 10 (exceptionally creative).

Since sample C comprised the three ethnic groups in Singapore (i.e., the Chinese, the Malays, and the Indians), the minimum and maximum ratings of the adaptor and innovator styles, the mean adaptor and innovator ratings, and the standard deviations for these specific subgroups are shown in Table 4.2.

Table 4.1 Adaptor and innovator ratings for samples A, B, and C

	N	Min.	Max.	Mean	SD
Sample A—USA	139				
Adaptor rating		1.0	10.0	4.6	2.2
Innovator rating		1.0	10.0	7.3	1.9
Sample B—Singapore	199				
Adaptor rating		1.0	10.0	4.8	1.9
Innovator rating		1.0	10.0	7.1	2.0
Sample C—Singapore (Chinese, Malay, and Indian ethnic groups)	185				
Adaptor rating		1.0	10.0	4.9	1.9
Innovator rating		1.0	10.0	7.3	2.0

Table 4.2 Adaptor and innovator ratings for sample C—Chinese, Malays, and Indians

	N	Min.	Max.	Mean	SD
Chinese	84				
Adaptor rating		1.0	9.0	4.9	1.7
Innovator rating		2.0	10.0	7.5	1.6
Malays	54				
Adaptor rating		1.0	10.0	5.4	1.9
Innovator rating		1.0	10.0	7.3	2.2
Indians	47				
Adaptor rating		1.0	9.0	4.5	2.2
Innovator rating		2.0	10.0	6.9	2.1

Table 4.3 Implicit perceptions of adaptor–innovator creativity (across all samples)

Sample	*N*	Mean	SD	*t*	*p*
A, B, and C	523				
Adaptor rating		4.85	2.03	−19.51	0.001
Innovator rating		7.28	1.99		

In Table 4.2, the mean rating for the adaptive style ranged from 4.5 to 5.4, while the mean rating for the innovative style ranged from 6.9 to 7.5. Just like Table 1.1, the innovator style received higher ratings for creativity. Only the Malay group had both adaptor and innovator styles receive ratings across the full continuum; that is, both the adaptor and innovator styles were rated as 1 (not at all creative) and 10 (exceptionally creative).

Next, *t*-tests were used to analyze the differences between mean ratings for adaptors and innovators for samples A, B, and C combined. Table 4.3 shows the *t*-test analysis of the mean ratings of the adaptor and innovator for the 523 participants from the three samples combined.

Table 4.3 shows the innovator style received a higher mean rating (7.28) than the adaptor style (4.85) of all the participants involved in this study ($n=523$). The difference in the mean ratings is statistically significant with $p=0.001$ ($p<0.05$).

The second analysis of the data involved the open-ended question where qualitative analysis was conducted. The method of coded data was employed to categorize the responses for the open-ended question (Huberman and Miles 1994). Responses from all participants were compiled and each response was assigned a code. A code was created as long as there was a minimum of two similar responses from each sample. For each code, the frequency of similar responses was noted. A "miscellaneous" category was set up to include responses that did not fit into any assigned codes. A total of 87 codes, including the "miscellaneous," category was formed. Table 4.4 shows the top categories (codes) from sample A and sample B.

In Table 4.4, the top categories accounted for 404 responses (45.2 %) out of a total of 879 responses. The top category from the American sample was "arts/artistic" which accounted for 10.2 % of all the responses, while this category accounted for only 4.1 % of the Singaporean sample. In sample B, the top categories accounted for 424 responses (48.3 %) out of a total of 871 responses. The top category was "new," which accounted for 11.4 % of all the responses. Both samples have "think out of the box" as the category with the second highest number of responses.

Since sample C consisted of the three ethnic groups, a breakdown of categories from each ethnic group is provided in Table 4.5.

In Table 4.5, the top category for the Chinese group is "think outside the box," which accounted for 13 % of all the responses. The top category for the Malay group was "arts/artistic" which accounted for 9.3 % of all the responses, while the top category for the Indian group was "new," which accounted for 8.8 % of all the responses. Also, the Chinese had two categories, "bold" and "interesting," which were absent from the Malay and Indian samples. In all the three ethnic groups, a new category, "abnormal/weird," was formed. This category was absent in sample A (USA) and sample B (Singapore).

Table 4.4 Top categories reported from sample A and sample B

Sample	Category	Frequency	%
Sample A—USA (n = 139)			
Total number of responses = 879	1. Arts/artistic	90	10.2
	2. Think outside the box	50	5.6
	3. New	40	4.5
	4. Open	32	3.6
	5. Intelligent	30	3.4
	6. Problem solver	27	3.0
	8. Imagination	27	3.0
	10. Unusual	27	3.0
	11. Different	21	2.3
	12. Innovative	20	2.2
	14. Flexible	20	2.2
	16. Unique	20	2.2
Total		404	45.9
Sample B—Singapore (n = 199)			
Total number of responses = 871	1. New	100	11.4
	2. Think outside the box	60	6.8
	3. Innovative	43	4.9
	5. Different	43	4.9
	6. Unusual	41	4.7
	7. Arts/artistic	36	4.1
	9. Ideas	36	4.1
	10. Problem solver	22	2.5
	12. Bold	22	2.5
	13. Imagination	21	2.4
Total		424	48.6

Discussion

A clear pattern that emerged from the mean ratings of the adaptor and innovator was that the participants in the USA and Singapore indicated an implicit belief that a high level of creativity was more associated with Kirton's (1976) innovative style of creativity. There was a consistent higher mean rating to the innovator than the adaptor. If generalizable, there seemed to be a perceptual bias toward the innovator being more creative than the adaptor. This is in contrast with Kirton's (1976) explicit theory where it states that adaptors are equally creative as innovators, at least with regard to laypeople.

Studies by Puccio and Chimento (2001), Gonzalez (2003), and Muneyoshi and Kagawa (2004) noted similar findings where the innovator was rated as more creative than the adaptor. Puccio and Chimento (2001) believed that culture could have played a role in influencing the perception of the innovator style as being more creative since "innovation" is highly valued, marketed, publicized, and sought after. Furthermore, they suggested that "the popular phrase often used to describe creativity, 'out-of-the-box-thinking', seems to reflect a bias towards the paradigm-breaking style associated with Kirton's innovator" (p. 679).

Table 4.5 Top categories reported from each ethnic group in sample C (Singapore)

Sample	Category	Frequency	%
Chinese ($n=84$)			
Total number of responses=415	1. Think outside the box	54	13.0
	2. New	53	12.7
	3. Unusual	24	5.7
	4. Innovative	22	5.3
	5. Unique	17	4.0
	7. Different	17	4.0
	8. Problem solver	15	3.6
	10. Bold	15	3.6
	11. Arts/artistic	11	2.6
	13. Interesting	11	2.6
	14. Abnormal/weird	9	2.1
Total		248	59.7
Malays ($n=54$)			
Total number of responses=299	1. Arts/artistic	28	9.3
	2. Think outside the box	18	6.0
	3. Unique	17	5.6
	4. New	11	3.6
	5. Innovative	10	3.3
	7. Different	10	3.3
	8. Abnormal/weird	9	3.0
	9. Imagination	8	2.6
	11. Intelligent	8	2.6
Total		119	39.7
Indians ($n=47$)			
Total number of responses=259	1. New	23	8.8
	2. Think outside the box	21	8.1
	3. Innovative	17	6.5
	4. Unique	15	5.7
	5. Unusual	14	5.4
	7. Different	14	5.4
	8. Imagination	10	3.8
	9. Arts/artistic	7	2.7
	11. Problem solver	7	2.7
	13. Abnormal/weird	7	2.7
Total		135	52.1

Another possible explanation is that in the case of the USA, Western values on creativity were dominated by American ideology, whereby creativity was viewed as creating new and useful objects and ideas that significantly departed from existing ones (Weiner 2000). Also, because of a strong emphasis on freedom of expression, individualism, and democracy as reinforced by a political system that protected freedom and protesting rights, Americans were imbibed in a culture where they were encouraged to go beyond the existing frontiers (Weiner 2000). This implies that breaking paradigms and questioning the norms are hallmarks of a creative society, and these in turn seem to reflect the innovator style of creativity.

A similar situation was also found in the Singaporean samples. Although Singapore is an Asian country that is considered to be more collectivist in nature, where there is communal-based regulation of society (Hofstede 1984), the innovator style was deemed more creative, not unlike the participants in the American sample. Even within the national culture of Singapore, the three ethnic groups registered a similar implicit belief that high creativity was associated with the innovative style of creativity.

When laypeople from the USA and Singapore were asked to define creativity in their own words, one similarity between the samples was that most of the top categories of responses seemed to have an innovator bias in the laypeople's implicit theory of creativity. For example, words like (a) think out of the box, (b) new, (c) innovative, (d) unusual, and (e) different were some of the top categories from each national culture. This finding further corroborated the participants' implicit belief that creativity was more associated with the innovative style of creativity rather than the adaptive style.

On a larger scale, the participants did not share the explicit notion of what creativity is. The general consensus in the field is that creativity includes not only the features of novelty or originality but usefulness (appropriateness) as well (Amabile 1983; Mayer 1999). However, in the implicit theories of laypeople in this study, the concept of "useful" was clearly absent in their responses, even in the sample from the USA, a Western culture. There is a possibility that laypeople's implicit views were based solely on novelty.

Some differences were also noted. The top category from the American sample was "arts/artistic," unlike the Singaporean sample. Leung et al. (2004) noted, "In the West, creativity is often viewed as an individual activity, and that may be why creativity is typically associated with artists or scientists" (p. 121). This finding is consistent with Runco's observation (1999) that creativity is only found in the arts domain. Further, within one national culture, the Malay ethnic group equated creativity with the arts, unlike the Chinese and Indian ethnic groups.

Most of the categories from the three ethnic groups indicated a perceptual bias toward the innovator style. One category that seemed to be absent from the two national cultures of the USA and Singapore was "abnormal/weird." Words in this category included (a) crazy, (b) irrational, (c) eccentric, and (d) wacky. Thus, creativity was associated with ideas, behaviors, or products that were out of the norm or particular paradigm. This again reinforced the idea that creativity was more associated with the innovative style of creativity rather than the adaptive style.

This study confirmed other research studies using Kirton's explicit theory of adaption and innovation to access laypeople's implicit theories of creativity (Gonzalez 2003; Muneyoshi and Kagawa 2004; Puccio and Chimento 2001). Three very distinct cultures—Latin, Anglo-Saxon, and Asian—consistently gave higher scores to the innovator, an indication of a perceptual bias across various types of cultures toward the innovator style of creativity, which was in direct contention to Kirton's theoretical position.

A Western or Eastern culture is not entirely homogeneous. These are very broad terms that do not allude to a myriad of subcultures within a particular national culture.

The vast historical and sociopolitical differences in the Western and Eastern cultures simply do not justify treating these groups as uniform entities. Thus, the findings from this study imply that research in Western and Eastern conceptions of creativity should give way to more research within a particular national culture so as to unearth the richness of how creativity is conceived in various subcultures within a larger entity. Perhaps, instead of coming up with a common definition of creativity that can cross all cultures, the complexities of how creativity is conceived in various cultures should be recognized.

In conclusion, an explicit theory cannot be assumed to have a shared global understanding of its concepts and ideas. The findings from this study can pave the way for more research on implicit theories of creativity, where there can be a deeper appreciation of how creativity is viewed all over the globe. Any explicit theory on a psychological construct can incorporate testing it on the general population by way of implicit theories so as to add more rigor and acceptance within a given society.

References

Amabile, T. M. (1983). The social psychology of creativity: A componential conceptualization. *Journal of Personality and Social Psychology, 45*, 357–376.

Csikszentmihalyi, M. (1990). The domain of creativity. In M. A. Runco & R. S. Albert (Eds.), *Theories of creativity* (pp. 190–212). Newbury Park, CA: Sage Publications.

Davis, G. A. (2004). *Creativity is forever*. Dubuque, IA: Kendall/Hunt Publishing Company.

De Bono, E. (1971). *Lateral thinking for management*. New York: McGraw-Hill.

Glaveanu, V. D. (2014). *Thinking through creativity and culture: Toward an integrated model*. New Brunswick, NJ: Transaction Publishers.

Gonzalez, M. (2003) *Implicit theories of creativity across cultures*. Unpublished Master's Thesis. Buffalo, NY: State University College at Buffalo, Center for Studies in Creativity.

Guilford, J. P. (1950). Creativity. *American Psychologist, 5*, 444–454.

Hofstede, G. (1984). *Culture's consequences: International differences in work-related values*. Beverly Hills, CA: Sage Publications.

Huberman, A. M., & Miles, M. B. (1994). Data management and analysis methods. In N. K. Denzin & Y. S. Lincoln (Eds.), *Handbook of qualitative research* (pp. 428–444). Thousand Oaks, CA: Sage Publications.

Kim, K. H. (2005). Learning from each other: creativity in East Asian and American education. *Creativity Research Journal, 17*(4), 337–347.

Kirton, M. J. (1976). Adaptors and innovators: a description and measure. *Journal of Applied Psychology, 61*, 622–629.

Kirton, M. J. (1994). *Adaptors and innovators*. London: International Thomson Press.

Kirton, M. J. (1999). Adaptors and innovators: a description and measure. In G. J. Puccio & M. C. Murdock (Eds.), *Creativity assessment: Reading and resources* (pp. 121–131). Buffalo, NY: Creative Education Foundation.

Kroeber, A. L. (1944). *Configurations of culture growth*. Berkeley, CA: University of California Press.

Leung, K., Au, A., & Leung, B. W. C. (2004). Creativity and innovation: East-West comparisons with an emphasis on Chinese societies. In S. Lau, A. N. N. Hui, & G. Y. C. Ng (Eds.), *Creativity: When east meets west* (pp. 113–135). Singapore: World Scientific Publishing.

Li, J. (1997). Creativity in horizontal and vertical domains. *Creativity Research Journal, 10*(2-3), 107–132.

MacKinnon, D. W. (1965). Personality and the realization of creative potential. *American Psychologist, 20*, 273–281.
Martinsen, O., & Kaufmann, G. (1999). Cognitive style and creativity. In M. A. Runco & S. R. Pritzer (Eds.), *Encyclopedia of creativity* (Vol. 1, pp. 273–282). San Diego, CA: Academic Press.
Matsumoto, D. (1994). *People: Psychology from a cultural perspective*. Pacific Grove, CA: Brooks/Cole Publishing.
Mayer, R. E. (1999). Fifty years of creativity research. In R. J. Sternberg (Ed.), *Handbook of creativity* (pp. 449–460). New York: Cambridge University Press.
Muneyoshi, H., & Kagawa, Y. (2004). *Implicit versus explicit theories of creativity: East versus west. Poster session presented at the 6th student research and creativity celebration*, State University College at Buffalo.
Osborn, A. F. (1953). *Applied imagination: Principles and procedures of creative thinking*. New York: Scribners.
Puccio, G. J., & Chimento, M. D. (2001). Implicit theories of creativity: Laypersons' perceptions of the creativity of adaptors and innovators. *Perceptual and Motor Skills, 92*, 675–681.
Rudowicz, E. (2003). Creativity and culture: a two way interaction. *Scandinavian Journal of Educational Research, 47*(3), 273–290.
Runco, M. A. (1990). Implicit theories and ideational creativity. In M. A. Runco & R. S. Albert (Eds.), *Theories of creativity* (pp. 234–252). Newbury Park, CA: Sage Publications.
Runco, M. A. (1999). Implicit theories. In M. A. Runco & S. R. Pritzker (Eds.), *Encyclopedia of creativity* (Vol. 2, pp. 27–30). San Diego, CA: Academic Press.
Simonton, D. K. (2000). Creativity – cognitive, personal, and social aspects. *American Psychologist, 55*(1), 151–158.
Soh, K. C. (1999). East-west difference in views on creativity: Is Howard Gardner correct? Yes and no. *The Journal of Creative Behavior, 33*(2), 112–125.
Weiner, R. P. (2000). *Creativity and beyond: Cultures, values, and change*. Albany, NY: State University of New York Press.

Chapter 5
Group Creativity and Individual Creativity: A Case Study of the Differences Between Japanese and Chinese Creativity

Fangqi Xu

Abstract Any group consists of individuals. Once a person becomes a member of a group, however, his/her creativity will get affected under the influence of the social and cultural environment. This chapter compares and discusses individual and group creativity in China and Japan.

Introduction

As J. P. Guilford (1962) pointed out, "In this narrow sense, creativity refers to the abilities that are most characteristic of creative people" (p. 152); E.P. Torrance defined creativity as "the process of forming ideas or hypotheses, testing hypotheses, and communicating the results" (cited in Guilford (1962), p. 32); and Teresa M. Amabile (1998) argued that creativity is composed of domain knowledge, creative thinking, and motivation; whether creativity is the abilities, the process, or those three components, it is certain that it is one of the main human psychological characteristics.

When such an argument is made, there may be some readers who would worry about whether a group has creativity or not. Any group consists of individuals. Once a person becomes a member of a group, however, his/her creativity will get affected under the influence of the environment. If the environment is conducive, that is, if the environment promotes individual creativity, it produces the effect of $1 + 1 > 2$. On the other hand, if the environment is not good, that is, if the environment inhibits individual creativity, the result turns out to be $1 + 1 < 2$.

I am a native Chinese and have been living in Japan for 24 years. I am also a scholar in the field of creativity research. So, I have been researching what differences

F. Xu, Ph.D. (✉)
The Institute for Crerative Management and Innovation, Kindai University, Higashi-Osaka, Japan
e-mail: fangqijo@yahoo.co.jp

I.N. Dubina, E.G. Carayannis (eds.), *Creativity, Innovation, and Entrepreneurship Across Cultures*, Innovation, Technology, and Knowledge Management, DOI 10.1007/978-1-4939-3261-0_5

exist between Japanese and Chinese creativity. As a result, I have made some interesting discoveries. In the case of the Japanese, a group tends to enhance individual creativity; in the case of the Chinese, however, a group is apt to suppress individual creativity. There are two main reasons. First, with the exception of very talented individuals, generally speaking, individual creativity of the Japanese is weaker than that of the Chinese. Second, there is a Japanese tradition that with the inclination among Japanese people to underestimate individual abilities, they attempt to compensate the lack of individual abilities with group efforts. This is the so-called one hundred ordinary people are better than one genius. On the other hand, China has a tradition that with the tendency to overestimate individual abilities, they are likely to compete on their own.

This chapter sets up and discusses the following questions: "Why is Japanese group creativity stronger while individual one is weaker?" and "Why is Chinese individual creativity stronger although it becomes weaker once an individual joins a group?"

Individual Creativity and Group Creativity

My definition of creativity is made up of the three elements: the ability to generate new things or new value, the process to achieve it, and the product as a result. There are three keywords in this definition. They are "ability," "process," and "product." The ability means the individual talent in order to create new things or new value. In other words, if a thing generated by someone already exists or he/she is unable to create new value, it is not considered as creativity. It should be called "imitation." Imitation may be the entrance of creation, but must not be equated with creativity.

In addition, except a rare case (e.g.,, the one where someone gets a serendipitous encounter and comes up with some idea which will lead to an invention or a discovery later), it takes some time to produce new things or new value. Because while doing an activity like this (creative activity), you begin to think creatively and eventually produce an outcome (creative result), the process is very important and so should not be neglected. Many creativity scholars focus on the "process" part, too (Parnes 1962; Stein 1974).

The most common method for evaluating creativity is the assessment of the product. There is a wide variety of creative products in the world, so it is impossible to provide a clear answer to what kind of creative products exist. In the case of businesses, creative products may be as follows: the development of a new product or service, the improvement of a production process, the planning of a new promotional technique, the generation of a communication tool, the introduction of a new system for effective personnel evaluation, and so on. However, when we evaluate creativity that do not directly contribute to sales, for example, the personnel, the labor, the planning, and the general affairs, it is not enough just to evaluate the result. In such a case, it is necessary to evaluate creative processes and motivation.

Japanese Creativity: Based on Group Consciousness

A School Sports Day

If you compare Chinese and Japanese, you will find that if a Chinese person competed with a Japanese person, the former would win most of the time. But when a group of Chinese people competed with that of Japanese ones, almost always the Japanese group would win. Why is this? In a nutshell, Chinese has stronger individual consciousness, whereas Japanese has stronger collective consciousness. Now, why is Japanese collective consciousness strong? How is such consciousness trained?

From here, I will answer those questions through the case of my daughter. She came to Japan when she was 4 years old. Shortly after arriving in Japan, she entered a private nursery school and then a private kindergarten 6 months later. There were not so many children in either of the facilities, but both had good education programs. They had a school play and a sports day. I attended such events with my wife every year.

When I watched my daughter's sports day, I found that they did not have any individual events. Moreover, there were only two teams: the red team and the white team. The group which won the games shouted "We won!" in a loud voice. On the other hand, the group which lost the games cried "We lost!" hand in hand. Then, I thought this was just their own way.

Two years later, my daughter entered a public primary school in our neighborhood. The school had many students and held a sports day every year. Needless to say, I visited the event with my wife. However, when each class competed, I found that they had no individual events and, again, had only the red team and the white team although the school had about 600 students. The winning team shouted "We won!" in a loud voice, and the other group cried "We lost!" in a circle. The view was exactly as the same as those sports days of the kindergarten. Moreover, there were even some parents who were crying as their children's team lost. I thought it was not the school's original way, but it might be a common phenomenon among Japanese elementary schools. So, when other elementary schools held their sports day, I went to watch them. Consequently, I found that their ways of having a sports day was just the same. Then I thought it was the way that only kindergartens and elementary schools adopted.

Six years later, my daughter entered a public middle school in the neighborhood. When the school held its sports day, I made an effort to go and watch the event for my daughter as much as possible. To my surprise, there were no individual events again. The students were only competing in some team events such as relay races and group gymnastics. Team colors were already familiar, red and white. The team which won the games shouted "We won!" in a loud voice, and the other group cried "We lost!" in a circle and consoled each other. Because Japan has the 9-year compulsory education system, I suspected that only the schools included in the compulsory education held a sports day in the way they did. But I was wrong.

Three years later, my daughter entered a famous public high school after passing an entrance examination. I had been too busy to visit my daughter's sports days at this point. Asking my wife to videotape the events, I watched them at home on weekends. Because the school was famous, the number of its students was large, about 1000 students enrolled. However, their games, which they had designed by themselves, were made up of only team events such as a makeup dance, a piggyback-style cavalry battle, a tug-of-war, and so on. No individual event. This time only the number of teams increased. Up to the middle school, there were two teams on a sports day. This high school doubled the number of them. They were blue, red, yellow, and green with each color team made up of two classes. In other words each two classes competed with the other six. The resulting scene was the same as what was seen at the middle school. The team which won happily shouted "We won!" The others cried for losing games, vowing to win the next year.

A Classroom

I found the same phenomenon happening in the classroom. School teachers from elementary to high school do not usually rank their students on their report cards. At elementary school, the assessment of a student's performance is qualitative rather than quantitative. For example, my daughter's teacher assessed students' performances with "very good," "good," and "a little more" (i.e., under average). From middle school to high school even now, many schools evaluate students' performance with "excellent," "good," "fair," and "failure." After all, students cannot know how much difference there is between their own academic achievement and others'. As is often the case with other countries, people sometimes say, "He was the top in the class when he graduated." In Japan, however, this is hardly heard.

Through my observation of Japanese schools, one thing is now clear. That is, Japanese teachers from kindergarten to high school have been training their students to have collective consciousness very hard. At the same time, they have consciously suppressed their students' individual consciousness. How do adults act who were trained in such a way in their childhood? They often underestimate or ignore an individual and even themselves. Also, they regard it as vital that everyone acts as a member of a group and cooperates with each other to overcome difficulty.

Findings

At this point, I decided to make academic research into the following three questions: why Japanese people have very strong collective consciousness; whether it is related to Japanese history, culture, and/or tradition; and whether their collective consciousness is associated with Japanese group creativity. As the result of my research by examining in various ways, I discovered the following causes (Xu 2006):

1. Relation to the natural environment
   As is well known, Japan is an island nation. Its narrow land surrounded by the sea, with the frequent occurrences of natural disasters like earthquakes, tsunami, typhoons, and volcano eruptions, the country is hard to live in. Therefore, once a natural disaster happens, its villagers or townspeople have to fight with nature with everyone's power.
2. Relation to the industrial structure
   Rice cultivation and fishery were the only means of supporting Japanese's life a long time ago. It requires water to grow rice. Needless to say, while water must be fresh, the water resources have been limited. It is not so difficult to imagine that disputes over water occurred between villages. When such disputes arose, it required everyone who lived in the same village to unite in fighting. As for fishery, once a boat meets rough weather such as a storm or high waves, the boat would avoid sinking with everyone aboard working in unity.
3. Relation to the political system
   The political system here mainly means Japan's emperor system. The system is hereditary. Historically speaking, with the exception of the Edo Period (1603–1867), when so-called shoguns (samurai military rulers) kept a total control over Japan, the Emperor was sometimes the highest authority of the nation, and other times he/she was esteemed as the symbol of the country. In particular, the Japanese worshiped the Emperor as God until its defeat of 1945 in World War II. Under such overwhelming authority, people even with excellent abilities started to feel in their childhood that their own abilities were limited and to be obedient to authority. In other words, they were inclined to lose the confidence in their own abilities and the desire to enhance them.

So, where is it possible to see the relation between Japanese collective consciousness and their group creativity? Such a relation can be seen in the two facts. To put it simply, we can see such a relation in the fact that Japanese collective consciousness gives priority to group interests over personal ones. Also the relation exists in the fact that Japanese work in full force in cooperation with each other. These phenomena can be seen clearly, particularly in Japanese businesspeople who work in a company. Suppose that a project team for new product development was formed. All the members would begin to work hard whether it is working hours or not. Sometimes, the project members even purchase materials, resources, and/or tools out of their own pocket in order to create a prototype at home. The same phenomenon can be seen in the field of service as well.

It should be emphasized here that these phenomena are not in a small number in Japan. Here are several examples among many: Nissin Food's "Cup Noodle" (NHK 2002), Fuji Heavy's "Subaru" car (NHK 2011), Panasonic's "Home Bakery" (Nonaka 1990), Canon's "Mini Copier" (Nonaka 1990), Asahi Beer's "Super Dry" beer (Nonaka and Takeuchi 1995), Honda's "City" car (Nonaka and Takeuchi 1995), Suntory's "Iemon" green tea (Mine 2006), and so on. All these successes were the results of group creativity.

Another aspect of the priority of group interests over personal interests is how Japanese people value an outcome. If someone is working as a member of a group and gets a good result, he/she would not talk about his/her individual contribution so much but never forgets to acknowledge the group members' contribution. A typical example is Koichi Tanaka, who won a Nobel Prize in Chemistry and was called "the most famous office worker in the world" (Kuroda 2003, p. 9).

Koichi Tanaka studied engineering at Tohoku University and graduated in 1983. He got employed by Shimazu, which is an analytical instrument maker in Kyoto. He is said to have felt in those days that he would work just as an engineer until he retired. Fortunately, he was awarded the Nobel Prize in Chemistry in 2002 sharing the honor with John B. Fenn and Kurt Wüthrich "for the development of methods for identification and structure analyses of biological macromolecules" (quoted from the website of the Nobel Foundation: http://www.nobelprize.org/nobel_prizes/chemistry/laureates/2002/). Since Koichi Tanaka neither studied as a post-graduate student nor holds a Ph.D. degree, his position in the company was not high; at the time, his position was senior researcher, but he was appointed as Research Fellow by Shimadzu after winning a Nobel Prize.

Koichi Tanaka asked the Nobel Foundation to invite his four colleagues to attend the Nobel Prize Award Ceremony. Since there is no such precedent, his request was denied by the foundation. But because Tanaka repeatedly requested, the foundation finally invited his four colleagues to the ceremony on the condition that Shimazu would pay for their travel. Why did Tanaka care about this? The answer is in what he told at a press conference: "We five members conducted a series of those research studies together. If someone asks me who contributed most, it is not possible to give a clear answer (Kuroda 2003, p. 111)." When he spoke at the Nobel Prize Award Ceremony, he thanked each of his four colleagues by referring to their names. The press extolled his humility, collective consciousness, and disinterest in honor and status as the virtue of Japanese.

Conversely, it is difficult for someone with an outstanding achievement to be accepted in the Japanese society if he/she has strong individual consciousness. Shuji Nakamura is a typical example.

The university where Shuji Nakamura studied and got a master's degree was not particularly prestigious. He was employed by Nichia, a medium-sized local company.[1] In spite of these, after being assigned to the Development Division, Shuji Nakamura started to work on a global problem alone. In 1993, finally, he succeeded in developing a "high-brightness blue light-emitting diode (blue LED)," whose development was said to be impossible during the twentieth century. Because he developed green LED, white LED, and yellow LED one after another, the world paid attention to him as a promising candidate for the Nobel Prize.[2] Although Shuij Nakamura produced huge profits for the company with his LED inventions, his

[1] Now, Nichia is a big company for its sales of 309.7 billion yen and 8400 employees (2013 fiscal year), but it was a small company in 1979.

[2] Finally, Shuji Nakamura was awarded the Nobel Prize in Physics with Isamu Akasaki and Hiroshi Amano in October 2014.

company hardly gave him bonuses. Although he filed a lot of patents, he only got from Nichia a bonus of ¥20,000 (about US$200) per patent because his achievements were so-called service inventions. Shuji Nakamura was frustrated that his contribution was not appreciated by the company. In 1999, therefore, he left the company at last. The following year, Nichia sued Nakamura for patent infringement. On the other hand, Nakamura sued Nichia, demanding the company's payment to him of two billion yen (later he increased 20 billion yen) as the compensation for his LED inventions. On January 30, 2004, the Tokyo District Court recognized Nakamura's claim and ordered Nichia to pay the 20 billion yen, which gave a shock to the Japanese society. Nichia settled with Nakamura for 840 million yen in 2005.

If we regard this case merely as a battle for the compensation for invention between the employer and one technician, it does not have to be discussed here. What I have noted is how a distinguished creator (a solo player) was left out in the cold and was criticized in Japan. Although Shuji Nakamura had the outstanding achievement at Nichia, he could not get much promoted. He was only a development manager at the time of leaving.

In 1999, once Shuji Nakamura found himself unable to endure his situation anymore and so left the company, job offers were rushed in him from the USA. Ten prestigious American universities and five companies approached him. But no offers were from Japan. Stanford and UCLA were both prepared to provide a tenured professor position for Shuji Nakamura. After considering those offers, he joined the California State University, Santa Barbara, as the professor of Department of Engineering.

Why did not any Japanese universities and companies want Shuji Nakamura? The reason is that he stuck to his working style of solo play, ignoring team play which is the tradition of the Japanese society. Besides, after Nakamura sued Nichia in order to protect the interests of his own, he was criticized severely as being "greedy for money" and "a traitor." Judging from this case, it is clear that the Japanese society sees collective consciousness as being essential and suppresses individual consciousness. As the proverb goes, "A tall tree catches much wind." Japan is a country where individual creativity is hard to grow.

The Strength and Weakness of Collective Consciousness

Japanese people working very hard in a group work full of energy and increase his/her creativity as long as they belong to the group. All members being in this state, their group creativity is stronger than the simple sum of their individual creativities. Conversely, if someone does not belong to a group, he/she will be confined to their own world and miss nurturing their individual creativities. It is due to this that a group of Japanese is strong and a Japanese individual is weak. In a nutshell, Japanese creativity is based on their collective consciousness.

Chinese Creativity: Based on Individual Consciousness

In the Chinese society, individualism is the mainstream even though China is said to be a socialistic country. Basically its politics and business are individualistic as well. Over a long course of its history, many strong leaders appeared in each field of the society, but they did so with their individual power.

If you look at the situation of the Chinese who live abroad, you will know it clearly. Chinese individualism is true of not only old overseas Chinese who crossed the sea for finding a way of life a long time ago but also of the new overseas Chinese who went abroad for study after Chinese reform and opening began in the 1980s. Generally speaking, the educational background of overseas Chinese in the past was not high. However, they managed to survive by relying on their individual skills, especially a set of skills called the "three blades" (a kitchen knife for cooking, a scissor for clothes-making, a razor for a barber) and have lived a middle-class life. The new overseas Chinese have high educational backgrounds and a lot of them have jobs requiring special knowledge and skills in the country where they studied.

By the way, the USA is a country of individualism; thus, I think that individualism does not have anything to do with a country's political system, but it is associated with history and tradition.

As I did before for Japanese culture, I examined the roots of China's individualism from the perspectives of its natural environment, industrial structure, political system, school education, and "one-child policy."

1. Relation to the natural environment
   China is about 27 times greater in size than Japan. Therefore, there is vast space where people can go. Natural disasters such as typhoons, floods, droughts, earthquakes, mudslides, and so on frequently occur in China. Considering these two factors, running away from them is much simpler than struggling in vain with nature. So, the collective consciousness of Chinese people is weak; if any, it stays within the group of family.
2. Relation to the industrial structure
   Because China is a large country with the natural environment differing considerably by region, the industrial base is also different among the regions. For example, northern China has grasslands where nomadic tribes lived. In the middle, people have traditionally cultivated wheat, barley, or millet such as corn, sorghum, and sweet potatoes for a long time. These plants basically absorb the water from rainfall or irrigation systems. In the southern and coastal regions, rice is the staple food. That is why people have grown rice for a long time, although the crop does not have much importance in other regions in China. Therefore, even if there is a conflict between villages for securing water sources, it did not become a serious problem affecting life and death of the entire village as in Japan.
3. Relation to the political system
   The state of China has a long history over 4000 years. The imperial political system was adopted until 1910. Although an emperor's crown usually passed from him/her to the child, to the grandchild, and so on, it was not always hereditary. An emperor held the greatest power, but the power was sometimes lost to the lord of

another family name. The displaced emperor was doomed. Li Emperor of the Tang Dynasty (618–907), Zhao Emperor of the Song Dynasty (960–1279), and Zhu Emperor of the Ming Dynasty (1368–1644) are good examples here. In this case, regime changes were made by military force. Whoever won the decisive war became an emperor. In other words, there was a possibility that regardless of origin and identity, anyone could become an emperor if he/she was strong enough. In fact, in the Chinese history, there were many cases that a regime was overthrown by a peasants' rebellion. For example, Yuanzhang Zhu (1328–1398), who was born to a begging family, grew up to be leader of an uprising. Due to his brevity and victories, he rapidly expanded his influence. When he won the war with the government forces, he lowered the curtain of the Yuan Dynasty (1271–1368); established his own regime, namely, the Ming Dynasty; and became the first Ming Emperor.

4. Relation to the education
   Competitions among school children have become increasingly fierce in China. Previously, people called the college entrance examination as the "examination war," from which we can infer the intensity of the competitions. But in recent years, those competitions have already reached kindergarteners. Chinese buzzwords go, "If you do not join in the competition at kindergarten, you have already lost at the start line of your life." Therefore, parents are eager to invest in their child's education. For example, a lot of parents make their child learn ancient poems, so-called Olympic mathematics, foreign languages, musical instruments, sports, and so on. Meanwhile, they are looking forward to seeing the child grow as a genius. If a prodigy appears somewhere, or if a child passes the entrance examination of a high school or college with the highest score, he/she receives applauds from all circles in the country. The most unfortunate are those children who do not have the genius quality. After taking a lot of educational programs under the parents' request, they could not accomplish a significant result. After all, contrary to the expectations of parents, they end up dropping out of the competition. There are many cases where such children lost interest in learning altogether.

Under such a trend, teachers from kindergarten to college rank their students' performance each time they have a game or a test. Moreover, the teachers usually read the ranking in class so that everyone can know the results. In Japan, if a teacher reads out the names of the students near the bottom of the list, it may well be regarded as an invasion of privacy, but there is no such worry in China. In fact, this ranking announcement is the practice done in all subject areas. The logic behind this practice is that by doing it, teachers can encourage students near the top of the list to study hard in order to make the same level of achievement on the next test. Also it aims to notify the ones down the ladder on the tests that there is a huge difference between their performance and that of the rest and urge them to study harder in order to catch up. There are few parents who are uncomfortable with this practice in school. Rather than criticizing, the parents believe that it is needed for the purpose of knowing how their children are doing on tests. Suppose that a student took 95 points on a test and showed the result to the parent. In Japan, the parent would praise, "You did really well!" In China, however, the parent would say, "Study harder to take 100 points next time!"

Chinese students who receive such training grow up in one of the three ways.

First, as for the students who are used to this type of evaluation method and occupy the top of the class, their sense of competition becomes stronger and stronger. They study hard to get the top of the class with every subject. At the same time, their elitism also comes out little by little.

Second, there are students who are not accustomed to this type of assessment method and are near the bottom of the list. They have either of the following two tendencies. One tendency is that they are not necessarily dull, so they give up on the left-side-of-the-brain subjects and try to achieve something in the right-side-of-the-brain subjects. For example, some students put themselves in the world of sports or arts. The other tendency is about the students who have some learning disability and so their ranking will not go up. Although they try to do their best for their learning, they lose hope at early stage of their lives and quit school.

The third category includes the students who are in the middle of the ranking in the class. They are inclined to show one of the two behaviors. One is that if their ranking is closer to the top group, they study harder to catch up with them. On the other hand if they are closer to the bottom group, they tend to lose motivation to go up the ladder and so remain the status quo.

The three patterns of Chinese students growing up in school may not be an ideal situation in education, but this is the reality of the Chinese education. Many schools have a special class where there are only top students. This type of class can be seen from elementary schools to high schools. They call this class the "elite class." The schools put the best teachers and educational facilities to the class. As a result, the elite students, who account for only 20 % of the total, are using more educational resources than the rest of the students.

(5) Relation to the "one-child policy"

In the 1980s the Chinese government began to implement the national policy that one family can have only child in order to suppress the population increase. Consequently, most of the children who were born in urban areas are only child. Chinese media call them "after 80" (people who were born in the 1980s) and "after 90" (people who were born in the 1990s).

Those only children have never been looked after by or have taken care of sibling simply because they don't have any brothers and sisters. In many cases, they play alone, their individual consciousness is strong, and they have little awareness to cooperate with others.

What Japanese and Chinese Should Do?

Concluding this chapter, I will compare the advantages and disadvantages of Japanese collective consciousness and Chinese individual consciousness.

First of all, let's take a fictional case of a Japanese businessperson. He works for a Japanese company and one day he is told that it is going to transfer him to a branch office in a foreign country next month. What does he say? Of course he says, "Yes, Sir! No problem!" Once this person arrives in his new town, what he does first is to visit the Japan Chamber of Commerce and Industry or a local Japanese Association office. Why? Because he wants to satisfy the desire of belonging to a group as soon as possible. After he is accepted by the group as a new member, his motivation goes up and he begins working creatively. If not, he gets restless and is unable to work quickly. When he has achieved something, he is humble enough to avoid praising himself and never forgets to thank his colleagues who have supported him on the way. Therefore, his team members will like him more and more with his position in the group rising gradually. When his performance turns out to be poor, as long as he is aware that he can expect support by the group, he is willing to work harder. Once he comes away from the group, however, his motivation will be lost.

When we consider the same scenario with a Chinese businessperson, he would think about the eating at first. In other words, he would check commercial facilities such as restaurants, supermarkets, convenience stores, and so on and determine whether he would eat out or cook for himself every day. As the proverb goes, "Hunger breeds discontentment," there is a tradition of putting priority to food in China. Once he decides on all issues of his meals, he checks whether there is any rival or not. If there is a rival, his motivation increases and he gets more and more willing to compete. He does not have much interest in whether there is a Chinese Association in town. Because of his tendency of solo play, he thinks that such an association is of no use. In other words, his ability of environmental adaptation is better than that of the Japanese described above.

Generally speaking, a Chinese representative positions his work overseas as individual play. When the result goes up, he becomes more confident in his own ability. If the result is properly evaluated by the company, his willingness to take on even more challenging work will get up. On the other hand, if his performance is not very good, he will recognize his own incompetence. The possibility of leaving the company will be high if he is reprimanded by his boss for his low performance.

So, what should the Japanese and Chinese do respectively?

As for the Japanese, it is necessary for them to consider how to recognize their own abilities correctly. This means to identify objectively their strong abilities and weak capabilities. By exerting their good abilities, it is possible for them to be able to work better than others and so make better results. Thinking in that way makes it easier for their confidence to come out. When the confidence comes out, the consciousness to rely on a group becomes less and self-activity increases. It must be noted that I never deny such consciousness; rather, I would like to emphasize that the excessive consciousness that depends on a group should be changed.

In fact, there are many Japanese people whose creativity is superb in Japanese society. They include not only Nobel Prize winners, but also Japanese who are active worldwide, such as Seiji Ozawa (a conductor), Hayao Miyazaki (a film director), Tadao Ando (an architect), Junko Koshino and Ken Okuyama (designers), and so on.

Now, what is necessary for Chinese people?

Needless to say, what is needed most for the Chinese is that they strengthen collective consciousness and develop a sense of cooperation with others. In the world of business, solo play is important. With the intensity of competition and a limit to solo play, however, opportunities of group play will increase and so will its weight. The so-called project team is a typical form of group play.

In today's business world, globalization is accelerating. It is difficult for companies to conduct a specific business only in their own country. In other words, they must take advantage of not only in-house management resources, but also external management resources. Therefore, it is necessary to work in cooperation with others. If not, individual creativity will not be exerted.

Chinese companies are acting with the sense of crisis about this. In particular, competitive companies have focused on this problem quickly and are taking remedial measures. For example, Lenovo has been instructing its new employees "to fit the personal pursue to long-term development of the company" and emphasizes the "group spirit of cooperation and confidence in each other" (Lenovo Cultural Handbook 2002, p. 23). They recognize the groups which accomplish outstanding results every year.

Since the mid-1990s, the educational program called "Outward Development" began in China. The origin was the Outward Bound, which is a US-based outdoor education program and is popular there and abroad. We consider it Chinese version of Outward Bound with some improvements. The chief purposes are to make the participants recognize the necessity and importance of group activities, to make them experience the process of problem solving with a group, and to make them to feel the fun of group activities (sense of accomplishment). This program fits in the training of young people who has too strong individual consciousness. At first, the program was offered mainly by education and training institutions in the private sector but is being accepted by companies gradually. Now, the main practitioners are companies' human resources departments. For example, the leading Chinese enterprises, such as Haier, Lenovo, and Sany Heavy Industry, have integrated the program into their new employees' trainings. At the same time, Beijing University, Tsinghua University, and many business schools introduced the Outward Development as a part of their MBA and EMBA educational programs.

Acknowledgments The author would like to thank Hideki Muneyoshi, the Director of the Institute of Creativity Education in Japan, for his valuable comments.

References

Amabile, T. M. (1998, September–October). How to kill creativity. *Harvard Business Review*, pp. 76–87.

Guilford, J. P. (1962). Creativity: its measurement and development. In S. J. Parnes & H. F. Harding (Eds.), *A source book for creative thinking*. New York: Charles Scribner's Sons.

Kuroda, T. (2003). *Tanaka Kouichi's life style*. Tokyo: Daiwashobo (in Japanese).

Lenovo cultural handbook. (2002). Lenovo Human Resource Department.

Mine, N. (2006). *Why lemon did sold*. Tokyo: Subarusya (in Japanese).

NHK. (2002). *Project X: Magic noodle*. Tokyo: NHK Enterprise.

NHK. (2011). *Project X: The first my car in Japan*. Tokyo: NHK Enterprise.

Nonaka, I. (1990). *Management of knowledge creation*. Tokyo: Nihon Keizai Shinbun-sha (in Japanese).

Nonaka, I., & Takeuchi, H. (1995). *The knowledge-creating company*. New York: Oxford University Press.

Parnes, S.J. and Harding, H.F. (1962). A Source Book for Creative Thinking. New York: Charles Scribner's Sons.

Stein, M. I. (1974). *Stimulating creativity*. New York: Academic Press.

Xu, F. (2006). *The comparative research between Japanese and Chinese enterprises*. Kyoto: Nakanishiya (in Japanese).

Chapter 6
A Review on an Ancient Cultural Convergence: A Case Study on Arsacid Creativity and Inventiveness in Interrelationship with Their Greek Origin Ethnic Group

Safura Borumand

Abstract Flourishing process of information technologies (IT) and information-communication technologies (ICT), as well as the expansion of media communication during recent decades, overcame political borders and paved the way for tangible social and cultural communications which formed the transnational interrelationship far beyond political affairs of states. This issue contemporized with presenting theories on cultural convergence and its resourcefulness. The cultural history recognition of countries with old historical background, rich culture, and extensive geographical and ethnical sphere denotes that emphasizing on cultural convergence has been one of the influential policies in permanence and continuance of their political, social, and economic structure. The issue of Iran during the Arsacid (Parthian) era (247/8 B.C.E.–224 A.D.) is a cross section of creativity and inventiveness in cultural policy based on cultural convergence and tolerance. This chapter studies and introduces Arsacid's policy in creating cultural convergence with Greek-origin ethnic group and their subculture in Iran based on primary sources.

Historical Findings and Cases of Cultural Convergence

The expeditious development of IT and ICT through the recent decades has been a backdrop for emerging theories related to globalization (Castells 1999; Movius 2010) with an emphasis on the "culture" which propound the issue of cultural communications, transnational interrelationship, cultural convergence, homogenization, and hybridization (James 2006; James and Tulloch 2010; Pieterse 2003; Ghosh

---

S. Borumand, Ph.D. (✉)
History Department, Institute for Humanities and Cultural Studies,
64 Western Av, Kurdistan Highway, Tehran, Iran
e-mail: s.borumand@ihcs.ac.ir

I.N. Dubina, E.G. Carayannis (eds.), *Creativity, Innovation, and Entrepreneurship Across Cultures*, Innovation, Technology, and Knowledge Management, DOI 10.1007/978-1-4939-3261-0_6

2011; Kraidy 2005). The issue of cultural convergence, its methods, approaches, and devices can be studied in a new purview by observing the history of globalization and cultural history of the nations and countries with ancient historical background, rich culture, and extensive geographical and ethnical sphere. Sociocultural situation of Greek origin ethnic groups in Iran during the Arsacid (Parthian) era (247/8 B.C.E.–224 A.D.) which is influenced by the cultural policy of the government is an aptitude example of creativity and inventiveness in cultural convergence and tolerance. A review on the profile of Greco-Persian relationship until the Arsacid era will shed the light on the elements which construct and generate the phenomena of this cultural convergence.

The oldest traces of Greco-Persian mutual relationship dated back to the early Achaemenid period when Cyrus II (called "the Great") started his first expedition westward for conquering Lydia (for details see Herodotus, *Histories*, Book 1.75 ff.; Grayson 1975; Rollinger 2004; Lendering 2008). It has also to be mentioned that Medes had occupied Eastern Anatolia adjacent to the River Halys in 585 B.C.E. He dominated most of Western Anatolia by occupying the Lydian empire, and consequently Achaemenid western borders adjoined Greek cities on the Aegean coast and some of them, whom the Persians called *Yaunā* (Ionians), came under Persian rule in 547/46 B.C.E. Consequently, most of their habitants had to do military service or provide troops for Persian army. Meanwhile, Persian leaders neither had particular hostility toward the Greek cities nor had intention of changing their constitutional position in any fundamental way (Walser 1984, p. 14). In Anatolia, for instance, with its long history of local Greek colonization and influence, some Persian families had been well acquainted with the Greek language since Achaemenid times (Boyce and Grenet 1991, p. 371). Therefore, a bilateral relationship emerged between the Persians and Greek population of Anatolia in social and mercantile affairs, though it was interrupted due to the latter's confederation with the Western Anatolian Greeks and their Greek kinsmen in the Balkan Peninsula who were politically independent and unwilling for Persian domination. The conflict between Achaemenids and Greeks (in Anatolia and Balkan) culminated in nearly 10 years of warfare (490 until 480/79 B.C.E.) called "the Persian Wars or the Greco-Persian Wars." Due to the various dates which have been determined as the beginning of this event, the exact period covered by these terms is open to interpretation from 10 to 50 years. The maximum extent is known from 499 to 449 B.C.E.

Despite some expeditions and advancements in Greece and capturing Athens by the Persian army, the war ended by Greek's victories in Marathon, Salamis, and Plataea (Herodotus, *Histories*, Book V, VI, VII, VIII). Yet some dissension between Athens and Sparta created a new phase in Greco-Persian relationship which led to the conclusion of peace treaties such as the Kallias (Callias) treaty in 449 B.C.E. and three treaties between Sparta and Persia in 412/11 B.C.E. on the issue of the entire Aegean Sea and the strip of the Anatolian coast, which belonged to the League, as a restricted area from which Persian forces were excluded (for more details, see Badian 1987, pp. 1–39). Finally, as the result of the Peloponnesian War between Athens and Sparta, the Greeks of Anatolia were brought back under the Achaemenid rule. For the next few decades until Alexander's expedition from Macedonia toward the Greece and Persian Empire, the relationship between

Greeks and Persians continued in parley, while commercial and cultural interrelationship was inevitable (Miller 2002; Sarton 1966). Through this period, in one stage, about over 300 Greeks such as the Athenian statesman Themistocles, the Spartan fugitive king Demartus, the physician Ctesias, and Ionian and Sardian sculptors such as Telephanes communicated with the Achaemenid court and some militaries had employed in Achaemenian army (Herodotus, *Histories*, Book 7.101; Porada 1985; Burn 1985). Some similarities between the function of myths and gods in Persian and Greek religious ideology were another field for cultural transnational interrelationship and homogenization between Persians and Greeks (Shaki 2003).

The next phase of Greco-Persian relationship started when Philip II (359–336 B.C.E.), the king of Macedon, launched an expedition for conquering Greece and Persian Empire which unfinished when he was assassinated (Diodorus Siculus, *The Library of History*, 16.91–94). His successor Alexander III (336–323 BC), called "the Great," followed his footsteps, dominated Greece, and headed for an expedition toward the Persian Empire. As Diodorus Siculus mentioned in *The Library of History* (17-17.2–3):

> Alexander advanced with his army to the Hellespont and transported it from Europe to Asia. He personally sailed with sixty fighting ships to the Troad, where he flung his spear from the ship and fixed it in the ground, and then leapt ashore himself the first of the Macedonians, signifying that he received Asia from the gods as a spear-won prize. He visited the tombs of the heroes Achilles, Ajax, and the rest and honored them with offerings and other appropriate marks of respect, and then proceeded to make an accurate count of his accompanying forces. (See also Justin, *Epitome of the Philippic History of Pompeius Trogus*, 11.5.10–12.)

Diodorus Siculus continued by narrating that Alexander accompanied by the infantry counted 12,000 Macedonians; 7000 allies; 5000 mercenaries with 7000 Odrysians, Triballians, and Illyrians; and 1000 archers and Agrianians, which in all the infantry would be numbered about 30,000–32,000. There were about 1800 Macedonians, 1800 Thessalians, 900 Thracian scouts and Paeonians, and about 600 of the other Greeks, which gave the total number of 4500 cavalry. Alexander entered Asia with about 36,500 men (Ibid, 17.17.3–4). Other available sources also give account and list the Alexander's troop. Justin speaks of 32,000 foot and 4500 horse (11.6.2), while Plutarch lists 30,000–43,000 foot and 4000–5000 horse (*The Parallel Lives*, *The Life of Alexander*, 15.1). But in *De Fortuna aut Virtute Alexandri* (Vol. IV 1.3) he mentions that Aristobulus gave 30,000 foot and 4000 horse, Ptolemy 30,000 foot and 5000 horse, and Anaximenes 43,000 foot and 5500 horse. Arrian mentioned approximately 30,000 foot and 5000 horse (*Anabasis of Alexander*, 1.11.3).

The result of Alexander's expedition toward the east was three major battles in Granicus (334 B.C.E.), Issus (333 B.C.E.), and Gaugamela (331 B.C.E.) which conduced to conquer the Persian army (Diodorus Siculus, *The Library of History*, 17-64.1.6). Evidently much of Alexander's troops were killed and wounded (Ibid). Alexander entered Babylon (331 B.C.E.) and remained there for 30 days. A Greek garrison was also settled in Babylon (Arrian, *Anabasis of Alexander*, 3.16.4). Meanwhile, after Alexander's victories in Mesopotamia, many Greeks and Macedonians especially soldiers entered and migrated to old cities such as Babylon, Nippur, and Uruk freely or forcefully (Van der Spek 2009). Although Greeks and Macedonians came to these towns as triumphant, they had special characteristics which distinguished them as an ethnic group. Like any other social minority, they could be modified within a larger society that displays a unique set of cultural traits. Melvin Tumin's definition of ethnic group more specifically coincides with these newcomers in the old cities of Mesopotamia and Persia, as he mentioned ethnic group as "a social group which, within a larger cultural and social system, claims or is accorded special status in terms of a complex of traits (ethnic traits) which it exhibits or is believed to exhibit" (Tumin 1964).

According to *The Library of History* by Diodorus Siculus which had the detailed information of these events, when he marched out of Babylon and headed for Susa:

> There came to him, sent by Antipater, five hundred Macedonian cavalry and six thousand infantry, six hundred Thracian cavalry and three thousand five hundred Trallians, and from the Peloponnese four thousand infantry and little less than a thousand cavalry. From Macedonia also came fifty sons of the king's Friends sent by their fathers to serve as bodyguards. The king welcomed all of these, continued his march. (17-65.1.2)

Alexander's expedition continued eastward and after reaching to present-day Afghanistan, some of his soldiers refused to follow him. Diodorus relates that after the death of Alexander, 23,000 Greeks lived in that area and were willing to march back home (Ibid, 18. 7). However, they stayed there and established a Greek colony which a few years later had become a base for establishment of an important Greek colonial city during the Seleucid period and thereafter (during the reign of Antiochus I (r. 281–261 B.C.E.)). Its remains have been excavated in the northeast of Afghanistan located in Ai Khanoum (Tepe) in the Bactriana area which represents the characteristics of a Greek city and its culture (Bernard 1985; Martinez-Sève 2014). Indeed, one of the main characteristics of Alexander and the Seleucids is their aim for an extensive colonization program (Van der Spek 2009, p. 106). According to Plutarch (*De Fortuna Alexandri*, Vol. IV, 5.1), Alexander founded 70 towns which most of them probably included preexisting settlements, permanent military garrisons (*katoikiai*), and temporary military settlement (*phrouria*) which repopulated during Alexander's expedition. However, the actual number of cities directly founded by Alexander is known to be at least 20 which most of those took his name and were called "Alexandria." One of these cities was Alexandria on the Tigris or Charax which was located on the northwest coast of the Persian Gulf (located now in Khorramshahr and Basra) probably in order to serve as commercial base and as a port for Babylon which related this city to the maritime activities in the Persian Gulf and Indian Ocean. As Pliny (*Natural History*, VI, 138) says, its habitants were people from the royal city of Durine and veteran Greek and Macedonian soldiers who got a privileged position and a separate city quarter. As the sociologist, John Porter mentioned "the first ethnic group to come into previously unpopulated territory, as the effective possessor, has the most say. This group becomes the charter group of the society, and among the many privileges and prerogatives which it retains are decisions about what other groups are to be let in and what they will be permitted to do" (Porter 1965, p. 60). This definition corresponds with the privilege that Macedonians and Greeks had in new towns across Persia which most of them were located at the fringe of the main trade routes.

On their way back to Babylon from the east while crossing the desert, many of Alexander's soldiers died of hunger, lack of food, and length of his march (Diodorus Siculus, *The Library of History*, 17.105.6–8). Most of his soldiers were discontented. This event along with prolong of his endless and oppressive expeditions, must have been the reason that Alexander adopted a cultural policy after returning to Susa (February or March 324 B.C.E). According to Arrian of Nicomedia (*Anabasis of Alexander*, 7.4), Alexander forced many Macedonians to marry native women. As for all the Macedonians who had already married Asian women, Alexander ordered a list of their names to be drawn up; they numbered over 10,000, and Alexander offered them gifts for their wedding. Although nearly most of these marriages ended in divorce, it was a cultural policy for generating an ethnic integration and hybridization. Meanwhile, the viceroys from the newly built cities and the rest of the territory subdued in war came to Alexander,

> bringing with them youths just growing into manhood to the number of 30,000, all of the same age, whom Alexander called *Epigoni* (successors). They were accoutered with Macedonian arms, and exercised in military discipline after the Macedonian system. The arrival of these is said to have vexed the Macedonians, who thought that Alexander was contriving every means in his power to free himself from his previous need of their services. (Ibid, 7.6)

Furthermore, some of his companions and soldiers who were unfit for service on account of age or any other reason went back of their own accord, to the number of about 10,000. Alexander ordered that if any of them had children by Asiatic wives, they had to leave them behind:

> Lest they should introduce into Macedonia a cause of discord, taking with them children by foreign women who were of a different race from the children whom they had left behind at home born of Macedonian mothers. He promised to take care that they should be brought up as Macedonians, educating them not only in general matters but also in the art of war. He also undertook to lead them into Macedonia when they arrived at manhood, and hand them over to their fathers. (Ibid, 7.12)

What Arrian narrated is the emergence of hybrid generation of Greco-Persian origin in Iran during the last decades of the fourth century B.C.E. Building new cities (*poleis*) with mixed population and arranging mixed marriages between Greeks, Macedonians, i.e., Hellenes and Persians, were among the aims of Alexander for making a cosmopolis and unified world (Plutarch, *Moralia.*). But after Alexander died in 323 B.C.E., the dispute of his warlords and commissioners started a war which disintegrated his territory and ended in the victory of Seleucus I Nicator over the Mesopotamia within the eastern Persian frontiers. However, he followed Alexander's policy and established cities "throughout the entire length of his dominions and named 16 of them Antioch after his father, five Laodicea after his mother, nine after himself, and four after his wives, that is, three Apamea and one Stratonicea" (Appian, *History of Rome*: *the Syrian Wars*, 57). One of them was Seleucia on the Tigris (now in Tell Umar, about 30 km south of Baghdad) which he chose as his capital. Many of Greeks leaved Babylon and settled in that city (Pausanias, *Description of Greece*, 1.16.3). During Antiochus I viceroyship until he succeeded his father, Seleucus I, in 281 B.C.E., the Macedonians of Babylon who lived there since Alexander's period were deported to Seleucia in the Tigris. The city which is called "the city of Kingship" had about 600,000 inhabitants of different ethnic backgrounds such as Macedonians, Greeks, Babylonians, Syrians, and Jews (Van der Spek 2009, p. 106).

About more than a century later under Antiochus IV Epiphanes (175–164 B.C.E.), again a Greek community settled in Babylon which had the privilege of political status (*politeia*). The Babylonian *Chronicle concerning the Greek Community in Babylon* (Greek Community Chronicle; BCHP 14) is an important source which mentions this event. The Greeks and Babylonians had each their own central governments and institutions (Van der Spek 2009, p. 109), a social system which lasted about 100 years later until the Parthian period around 77 B.C.E. as long as the cuneiform records are available. There are some criteria about these Greeks. The question is whether they were veteran soldiers of different nationalities who spoke the language of command and recognized as Greeks or they may have originally been Greeks (Van der Spek 2005). According to archeological findings in Seleucia, Greeks had their own cultural life and a sense of community, oneness, or peoplehood which derives from emphasizing on a shared ancestry, common roots, and experiences or heritage. The ruins of a theater and gymnasium have been excavated which dated back to the Hellenistic period. A Stoic school was established in Babylon and the city became a center for Greek civilization (Strabo, *Geography*, 16.1.6; Plutarch, *The Parallel Lives*, *The Life of Lucullus*, 22.5). Another characteristic of Babylon as a multiethnic society was its hybridized inhabitants such as poet and philosopher Herdicus of Babylon whose available facts introduced him with a multiple identity as Greek and Babylonian (Van der Spek 2009, p. 110).

Meanwhile, like other multiethnic societies, there was some heterogeneity in that city. After the death of Antiochus IV, conflicts arose between Babylonians and Greeks due to some issues on landowning which ended by a court session (Van der Spek 2009, p. 111). Apart from those issues, "The Hellenes had neither incentive nor means to impose their idolatrous religion on sacerdotal Mazdeans and revealed Mazdaism, or a quasi-democratic way of life on absolutist Persians. The sole desire of the expatriate Greeks, who soon felt alienated from their homeland, was to adjust themselves to the new conditions, govern the land, and collect taxes. Hence, under their hegemony every aspect of Hellenization of the country developed gradually, fortuitously, and unevenly." (Shaki 2003).

The hybridized government of Seleucids lasted about 70 years in Iran. Finally, the revolt of Arsacids (Parthians) from the northeast of Iran in 247/8 B.C.E. resulted in the overthrow of Seleucids. Finally, the last Seleucid rulers' territory was bounded in the wall of Antioch, a Greek-based city on the eastern side of the Orontes River, near the modern city of Antakya, Turkey (100 B.C.E.). After about 70 years of Greeks' presence and residence in Iran, people who still were Greek, had Greek origin, or had been hybridized lived in mixed cities, still influenced by Greek culture, rituals, and traditions. They had their own career and communicate with native Persians, while the vast native urban, rural, and tribal population retained their identity while finding the positive similarities between Greek and Persian culture and appreciated them. As for government policy, the characteristic of the Arsacid government is summarized in two concepts: federalism and tolerance. The policy which was coincided with their vast realm comprised of several provinces, various ethnic groups with multifarious rites and customs "ranging from total Greek autonomy in Bactria, to Persian sovereignty in Fars" (Shaki 2003). While in the Seleucid period the Greeks seem to have made little attempt to learn Oriental languages, their subjects, in order to communicate with their new rulers, had to learn Greek. As Plutarch (*De Fortuna Alexandri*, Vol. IV, 1.5) narrated, due to the educational process which started from the Alexandrian period and conduced to teaching Greek language to children of nobility, "Homer was commonly read, and the children of the Persians, of the Susianians [citizens of Susa], and of the Gedrosians [Baluchis] learned to chant the tragedies of Sophocles and Euripides." This process had its effect through decades; Greek language and handwriting had been used during the Arsacid era along with Parthian Pahlavi/Aramaic as concurrent media of communication in mixed cities with Greco-Persian habitants. Legends on Arsacid (Parthian) coins are all in Greek alphabet which included Greek epithets. Although inscriptions gradually were scrawl and unreadable, only the king's name in Aramaic characters is found on the coins of the last Arsacid kings such as Mitradates IV; Vologases IV, V, and VI; Artabanus V; and Artavasdes (129–224 A.D.). Greek language was used as complement in Parthian Pahlavi reliefs and this procedure continued during Sasanian era.

Moreover, signs on the early coins of Mitradates I (r. 171–138 B.C.E.), the third Arsacid king, can be interpreted as an image of cultural convergence. On his first coins, Mitradates I is depicted seated on omphalos holding a bow in his hand. Omphalos was an important stone in Hellenic religious symbolism related to Greek myths such as Cronus and Apollo and depicted world centrality and power. Mitradates' image on these coins resembled the image of Apollo seated on omphalos with a bow in his hand which was previously depicted on some Seleucid coins. The image rendered Mitradates' domination over Seleucid power. Mitradates I "was largely responsible for the political expansion of Parthia" (Sellwood 1993, p. 281) and finally occupied Mesopotamia (c. 140 B.C.E.) "with its celebrated city of Seleucia in the Tigris" (Sellwood 1993, p. 282) which had long been populated by Greek-origin habitants and was the most important city that minted coins. As Mitradates I occupied Mesopotamia, the image of omphalos changed into the famous Persian throne; this new image speaks of his final victory over the Seleucids and shows the wisely used Greek and Persian symbols as a tool for representation of a political message to Greek-origin subjects of Arsacids territory, especially those who lived in Seleucia (Borumand 2014). Mitradates was the first Arsacid king who used the legend ΒΑΣΙΛΕΩΣ ΜΕΓΑΛΟΥ ΑΡΣΑΚΟΥ ΦΙΛΕΛΛΗΝΟΣ, "of the great king Arsaces, philhellene." He used epithet *Philhellene*, the one who admires Greece or Greek culture, as "a somewhat transparent attempt to placate the Greek commercial element in the newly conquered lands" (Sellwood 1993, p. 282). Few decades later Mitradates II used this epithet on coins minted in Susa while he was supplanted there by other kings around 94 B.C.E. aimed at ranging his Greek subjects among his supporters (Sellwood 1993, p. 285).

After Mitradates I conquered Seleucia on Tigris, Greeks and Macedonians continued to exist there and became ordinary citizens who cherished their identity. Due to the tolerance policy of Arsacids, "Greek cities like Seleucia kept their political institutions and language and the Greek community in Susa remained a distinctive group" (Van der Spek 2009, p. 106). This situation can justify another aspect of cultural convergence during the Arsacid era which its signs have been represented on their coins. Greek myths such as Zeus, Apollo, Heracles, and Dioscuri were depicted on coins minted in Seleucia on the Tigris. Arsacid kings' investiture scene was shown accompanied with Nike, the Greek goddess who personified victory, while other Greek deities such as Tyche, Demeter, and Artemis along with a various Greek mythological signs and symbols were depicted on Arsacid coins. Coins minted in the semi-independent province of Characene (aforementioned Charax was its main city) were also designed with Greek symbols, the epithet *Philhellene* and legends. Greek myths on Arsacid coins all had substitutes in Persian mythology which their depiction was unconventional in Persian culture. Therefore, by using Greek myths and symbols resembling their identical Persian ones, a pattern of cultural convergence would emerge. Zeus–Ohrmazd, Anāhita–Aphrodite, Mithra–Apollo, and Varahran–Heracles were identical myths which were worshipped and venerated by both Zoroastrian Persians and Greeks in multicultural and multiethnic societies throughout the Arsacid realm. Other aspects of cultural convergence have been visualized in Arsacid art. Various genres of visual arts had become the integration of Greco-Persian culture and aesthetic tradition (see Ghirshman 1962 for more details). Literature was another field of experiencing cultural interrelationship. Arsacid elites, nobles, and courtiers were well acquainted with Greek literature. The tragedy of *Bacchae* by Euripides was played for Orodes II (ca. 57 B.C.E.), the Arsacid king, when the messenger arrived with news of the Roman's defeat and the death of Crassus, the famous Roman general whose ideal was the revival of Alexander's empire and conquering Persia (Plutarch, *Life of Crassus*, XXXIII).

Conclusion

The issue of culture is linked to meaning, knowledge, talents, civilization, and values. Culture strengthens social correlation among communities and resulted in social cohesion which is a set of social values related to people with different backgrounds. On the other hand, creativity is the combination of cognitive elements such as the ability to connect ideas, to see similarities/differences, and to be unorthodox. A cultural-based creativity is originated from creative people with personal abilities to think imaginatively or metaphorically, to challenge the conventional, and to call on the symbolic communication. Other element for emerging cultural-based creativity is the social environment which encourages creativity. Cultural convergence is an aspect of cultural-based creativity which can be traced in countries with ancient historical background, rich culture, and extensive geographical and ethnical sphere. The issue of Greek-origin society in Arsacid territory and the correlation that resulted in, was a cross section of cultural-based creativity which can be defined as a forerunner of cultural convergence in the classic world.

During the early years of Arsacid's rule in Iran, Greek-origin society had been spread and planted in some important areas such as old cities of Mesopotamia, Persian capitals, cities with commercial importance, and some newly established cities which were located alongside trade routes. They had established an ethnic community and sociocultural institutions and played an indicative role in mercantile affairs. Furthermore, Arsacids had their own tribal background and tradition which respects every individual opinion. This context resulted in choosing a democratic policy. A federalized government and toleration toward ethnicities especially Greek-origin ethnic group and respecting their citizenship were the sequels of this policy. They realized and appreciated Greek inhabitants' entity, nationhood, and functions. Emphasizing on conformity, parallelism, and homogeneity of some cultural elements such as myths was an expedient intermediary for coalescence of the society and mass citizenry with Greek-origin ethnic group.

The importance of this attitude toward Greek habitants in Iran will be defined with a scrutiny on the endless wars between Arsacids and the Roman Empire who considers itself as the patron of Greeks and Greek culture. Nevertheless, the Greek ethnic group who enjoyed the Arsacid cultural policy stayed in Iran and was gradually absorbed in Persian society and culture. Therefore, the ability of Arsacids to recognize cultural and symbolic similarities between Persians and Greeks and be unorthodox resulted in a cultural-based creativity. It emerged as the culture of ethnic tolerance which was exercised through emphasizing on cultural convergence. This was a new and creative approach toward interrelationship in a multiethnic society that had never been experienced before in ancient societies. This cultural policy was the main characteristic of the Arsacid rule which resulted in the continuance of this dynasty for about 470 years.

References

Appian, *History of Rome: the Syrian Wars.*

Arrian, *Anabasis of Alexander*, 1, 3.

Badian, E. (1987). The peace of Callias. *The Journal of Hellenic Studies, 50*, 1–39.

Bernard, O. (1985). ĀY K̲ĀNOM, *Encyclopeadia Iranica*, III(2), pp. 124–126, Retrieved September 23,2014,fromhttp://www.iranicaonline.org/articles/ay-kanom-or-a-khanum-tepe-a-local-uzbek-name-lit

Borumand, S. (2014). Archer, omphalos, throne: a review on connotation of three symbols on Parthian coins. *Cultural History Studies, 20*, 1–14.

Boyce, M., & Grenet, F. (1991). *Handbuch der oreientalistik.* Leiden: Brill.

Burn, A. R. (1985). Persia and the Greeks. In *Cambridge history of Iran* (Vol. II, pp. 292–391). Cambridge: Cambridge University Press.

Castells, M. (1999, September). *Information Technology, Globalization and Social Development*, Unrised Discussion Paper No. 114, United Nation Research Institute for Social Development. Retrieved from http://www.unrisd.org/80256B3C005BCCF9/(httpAuxPages)/F270E0C066F3DE7780256B67005B728C/$file/dp114.pdf

Diodorus Siculus, *The Library of History,* 16, 17.

Ghirshman, R. (1962). *Persian art, The Parthian and Sasanian dynasties, 249 B.C.-A.D. 651.* New York: Golden Press.

Ghosh, B. (2011). Cultural changes in the era of globalization. *Journal of Developing Societies, 27*(2), 153–175.

Grayson, A. K. (1975). *Assyrian and Babylonian chronicles*. New York: J. J. Augustin.
Herodotus, *Histories*, Book I, V, VI, VII, VIII.
James, P. (2006). *Globalism, nationalism, tribalism, bringing theory back in*. London: Sage Publication.
James, P., & Tulloch, J. (2010). *Globalization and culture, Vol. 1: Globalizing communications*. London: Sage Publications.
Justin, *Eptiome of the Philippic History of Pompeius Trogus*, 11.
Kraidy, M. (2005). *Hybridity, or the cultural logic of globalization*. Philadelphia: Temple University Press.
Lendering, J. (2008). The end of Lydia: 547? Retrieved September 23, 2014, from http://www.livius.org/men-mh/mermnads/547.html
Martinez-Sève, L. (2014). The spatial organization of Ai Khanoum, a Greek city on Afghanistan. *American Journal of Archaeology, 118*(2), 267–283.
Miller, M. C. (2002). Greece. Greco-Persian cultural relations. *Encyclopaedia Iranica*, Vol. XI, Fasc. 3, pp. 301–319. Retrieved September 23, 2014, from http://www.iranicaonline.org/articles/greece-ii
Movius, L. (2010). Cultural globalisation and challenges to traditional communication theories. *PLATFORM: Journal of Media and Communication, 2*(1), 6–18.
Pausanias, *Description of Greece*, 1.
Pieterse, J. N. (2003). *Globalization and culture: Global melange*. Lanham, MD: Rowman & Littleeld.
Pliny, *Natural History*, VI.
Plutarch, *De Fortuna Alexandri*, Vol. IV.
Plutarch, *Life of Crassus*, XXXIII.
Plutarch, *Moralia*.
Plutarch, *The Parallel Lives, The Life of Alexander*.
Plutarch, *The Parallel Lives, The Life of Lucullus*.
Porada, E. (1985). Classic Achaemenian architecture and sculpture. In *Cambridge history of Iran* (Vol. II, pp. 793–827). Cambridge: Cambridge University Press.
Porter, J. (1965). *The vertical mosaic: An analysis of social class and power in Canada*. Toronto, ON: University of Toronto Press.
Rollinger, R. (2004). The median "empire", the end of Urartu, and Cyrus' campaign in 547. *Proceedings of the First International Conference on Ancient Cultural Relations between Iran and West Asia*, Tehran. Retrieved September 23, 2014, from http://www.achemenet.com/ressources/souspresse/annonces/Rollinger-Iran.pdf
Sarton, G. (1966). *A history of science: Ancient science through the golden age of Greece*. Cambridge, MA: Harvard University Press.
Sellwood, D. (1993). Parthian coins. In E. Yarshater (Ed.), *The Cambridge history of Iran* (The Seleucid, Parthian and Sasanian Periods, Vol. 3(1), pp. 279–298). Cambridge: Cambridge University Press.
Shaki, M. (2003). Greece. Greek influence in Persian thought, *Encyclopeadia Iranica, Vol. XI, Fasc. 3*, pp. 321–326. Retrieved September 23, 2014, from http://www.iranicaonline.org/articles/greece-iv
Strabo, *Geography*, 16.
Tumin, M. (1964). Ethnic group. In J. Gould & W. L. Kolb (Eds.), *A dictionary of the social sciences* (pp. 243–244). New York: Free Press.
Van der Spek, R. J. (Bert). (2005). *Greek community chronicle (BCHO 14)*. Retrieved September 16, 2014, from http://www.livius.org/cg-cm/chronicles/bchp-greeks/greeks_01.html
Van der Spek, R. J. (Bert). (2009). Multi-ethnicity and ethnic segregation in Hellenistic Babylon. In T. Derks & N. Roymans (Eds.), *Ethnic constructs in antiquity, the role of power and tradition (Amsterdam archaeological studies 13)* (pp. 101–115). Amsterdam: Amsterdam University Press.
Walser, G. (1984). *Hellas und Iran: Studien zu den griechisch-persischen Beziehungen vor Alexander*. Darmstadt: Wissenschaftliche Buchgesellschaft.

Chapter 7
A Relationship Between Creativity, Management Styles, Culture, and Language (A Case of Japanese and English)

Michihiko Esaki and Chris Nishihama

Abstract This chapter investigates the correlation between the Japanese and English writing formats and syntax with respect to management and creativity of the Japanese and Americans. As creativeness is known to be born from the lateral movements of brain activity, the authors of this chapter try to demonstrate that creativeness comes naturally from the syntax. They also discuss how management styles can be attributed to both the language and cultural differences.

Introduction

This chapter investigates the correlation between the Japanese and English writing formats and syntax with respect to management and creativity of the Japanese and Americans. The two writing styles in the Japanese language, one written vertically from right to left and another horizontally from left to right, play a role in finding the "main keyword" necessary to define the theme of any given project and the steps needed to carry them out as shown in the Advanced Project Management Methodology by Dr. Esaki (2009).

Additionally, the Japanese syntax is related to the social mind-set of the Japanese people. Individuality is often shunned upon by the Japanese since they prefer oneness with the group and any deviation leads to one's isolation from the rest of society. For example, in the sentence "I make you happy," the Japanese wording using English is "I you happy make." The pronouns "I" and "you" are found together, supporting the idea that "Neither I nor you are alone" and "We are doing this together." Looking into the brain hemispheres, "I" and "you" are governed by the left/right sides of the brain and

M. Esaki, Ph.D. (✉)
DTCN International, Gifu, Japan
e-mail: esaki@dtcn-wisdom.jp

C. Nishihama
Gifu, Japan
e-mail: jccompassion@yahoo.co.jp

I.N. Dubina, E.G. Carayannis (eds.), *Creativity, Innovation, and Entrepreneurship Across Cultures*, Innovation, Technology, and Knowledge Management, DOI 10.1007/978-1-4939-3261-0_7

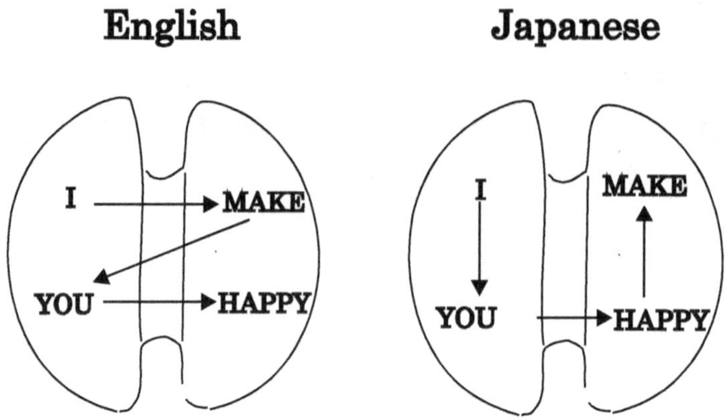

Fig. 7.1 English and Japanese word sequence causing different activity sequences of left and right sides of the brain. *Source: Authors' own work*

"happy" and "make" by the right side. As creativeness is known to be born from the lateral movements of brain activity, creativeness of Americans comes naturally from the English syntax which causes the brain to shift activity, in this case, from the left to right twice ("I make you happy," left to right to left to right). These shifts in activity also cause different eye, head, and body movements during the thought processes, but the same motion at the end result in both the Japanese and Americans.[1]

Historically, the Japanese have always been able to improve upon the ideas made by others (manufacturing, process), but had far fewer innovations (creativity) than the Americans. It is known that one of the requirements of creativity is to have lateral movement of activities in both the left and right hemispheres (Tsunoda 1985; Carter 2000; Shobe et al. 2009), which make Americans naturally better at creativity than the Japanese. The Japanese, however, have better aptitude at manufacturing due to the horizontal and vertical (matrix) viewing ability which comes from the Japanese writing formats. As we demonstrate in this chapter, management manufacturing styles can be also attributed to both the language and cultural differences.

Wordings

As we already noted in Introduction, in the English sentence, "I make you happy," the wording is such that the left and right hemispheres of the brain are both very active. The words "I" and "you" (nouns) are active in the left hemisphere, whereas "make" and "happy" (verb and adjective) are active in the right hemisphere: L (noun) → R (verb) → L (noun) → R (adjective) (Sereno 1998; Fiebach et al. 2002) (Fig. 7.1).

[1]More details are provided at http://dtcn-wisdom.jp/E-explanations/NODDING-and-PMD-relation.pdf.

The Japanese syntax *Watashi ha anata wo shiawase ni shimasu*, when worded in English, becomes "I you happy make." This leads to less cross-activities from the left to the right hemisphere: L (noun) → L (noun) → R (adjective) → R (verb).

In the Japanese syntax, the context is such that rather than "I" alone, "we" are doing this together. This mind-set where a group rather than an individual does the activities is deeply rooted in the Japanese culture and is even revealed in the Japanese grammar as such (Dunn et al. 2011). The Americans, however, are able to express individuality more so than the Japanese. The subject "I" will do all the work. This "oneness" or "togetherness" is known as "isshindoutai."

It also has a unique effect on quality assurance. Any poor workmanship by any of the workers reflects upon the company as a whole. It is not one worker to blame but the entire company for allowing such things to develop in the first place. This mind-set has its origins in the samurai code. Disgrace to the clan can lead to "hara-kiri," a form of public suicide by all the samurai members so that the family name would not be tarnished. A similar example of this may be likened to how punishment of the whole platoon in the US Army for the misdemeanor of one member. Having grown up in North America, one of the authors of this chapter came to understand by reading articles that there were many instances where line workers would not put forth their best effort as they viewed their poor workmanship to be the store or car dealer's problem.

Workplace

In the Japanese workplace, whether it be in education, government, or business, paperwork such as distributing notices, applications by potential new recruits, formal requests for a paid vacation, etc., must be viewed and signed by several individuals who are responsible for those forms before they are accepted. This system allows for a unified consensus by all so that if anything should go amiss, no one person will be blamed. This wanting to be a part of a group stems from the unwillingness of the Japanese to carry all the responsibilities on their own. Most organizations prefer to carry on what had been done previously and try to avoid new endeavors. While this kind of system may prevent individuals from taking the initiative to do something completely new, it does offer a safe haven where risks are shared by all and, depending upon the degree of severity, the highest-ranking worker will take the blame alone and leave the company. One field that does suffer from this mentality is research. Many researchers in Japan have a difficult time convincing their superiors of the direction that they would like to take, and so many of them find themselves travelling abroad to other universities or companies to continue their research. Thus, not only do the Japanese lack the occasion for creativity compared to Americans, but they also are hindered to invent or create at their workplace.

Likewise, the mentality of elementary and secondary education is said to be such that "all students should hold hands and cross the finish line together." What this means is that all students should be equal and no individual should show a difference from the rest of the group. Whether it be sports day events or academics, the Ministry of Science and Education strives for an average outcome of the end result.

Isshindoutai has its origins in sword making. It is said that the Japanese mind-set comes from the dawn of the samurai era but, more specifically, through perfecting the art of sword making. Sword making began sometime around the seventh to eighth century which also coincides with Buddhism entering Japan from China.

Before the sword making could begin, the craftsmen must be "cleansed" spiritually by the prayers of the Shinto monk. This cleansing allows "oneness" with God. Then, with three or four craftsmen working in unison, they fold and form the sword in eight stages and the final product can have up to a million layers. The important point here is to work together in unison with one mind and one direction, "isshindoutai."

Buddhism in Japan represents oneness with nature and the balance of forces. Being one with nature, it is important not to create waste in order to preserve resources. As Japan lacks many of the natural resources found in other countries, the Japanese over the centuries have become extremely careful in preserving resources. Efficiency and recycling promote less need. Thus, when looking at the many Japanese arts such as the tea ceremony, flower arrangement, calligraphy, bonsai, aikido, kendo, etc., minimal usage of energy, recycling, silence, and oneness with nature are the core themes.

Emotions, Gestures, and Traditions

The silence of the Japanese or lack of expressing their thoughts has caused much misunderstanding among the Americans. The Japanese language lacks many different ways of expressing bewilderment, love, beauty, etc. However, it must not be misunderstood as the Japanese lacking in feelings or emotions. Being married to a Japanese national and living in Japan for over 23 years, Chris (one of the authors of this chapter) can say with certainty that the Japanese are capable of as much feelings as their American counterparts. What is not clearly stated or expressed is supposed to be understood as such between the Japanese. This "reading between the lines" is a necessity if one is to have some form of relationship with the Japanese. Of course, a man will state his love for the woman he loves and vice versa, but it is nowhere near the number of times Americans convey their love for one another.

As well, gestures and eye, head, and body movements during communication and thought processes between the Japanese and Americans differ greatly. This is most likely due to different locations where the brain is active. For instance, the Japanese have the tendency to move their heads while keeping their eyes central to the range of sight when thinking. However, Americans tend to move their eyes sideways when thinking. Compared to English, the Japanese move their heads and/or bodies forward when verbs, adverbs, and adjectives are expressed (right brain activity seems to be connected to this movement in the Japanese), while Americans use their arms and shrug their shoulders which are not seen in the Japanese expressions. An interesting study (Tsunoda 1985) was conducted in the past where Americans

Table 7.1 A linear "left-to-right" process in manufacturing

Process No.	Input	Pre-assurance activity	Output	Review (Post-assurance activity)	Output approved
1	● →	● →	● →	●	
2	● →	● →	● →	●	
3	● →	● →	● →	●	
4	● →	● →	● →	●	

and Japanese were subject to many different types of sounds to see which areas of the brain became active. In the case of insect sounds, although it was heard in both ears, it led to only certain areas in one hemisphere of the brain to become active for the Japanese and the opposite hemisphere for the Americans. The same sounds probably elicited different emotional responses.

Thus, it may be beneficial to arrange instruments catered to certain ethnic groups rather than to have one music catered to the whole world. The arrangement or positioning of certain musical instruments may stir different emotional responses in people of different countries which in turn may increase musical enjoyment and appreciation.

Table 7.1 depicts the process steps in manufacturing showing the relationship between input and output and its quality inspection.

The management of ordering materials, processing, timing the preparation of lines, quality check, etc., are critical in creating an efficient manufacturing process. The vertical and horizontal writing formats of the Japanese language play a role in being able to see at a glance the bottlenecks, problem areas, key steps, etc. The speed of picking up the important points of such tables by the Japanese is not easily matched by the average North American, as it is much like reading a Japanese book. The theme or reasons for the process steps can be understood if one views the previous process. Thus, if one can understand the process steps vertically, one will be able to find the key steps in maintaining or improving the quality of the end product (Esaki 2009).

Being able to see the interrelationship of the process steps as a whole has its benefits not only in manufacturing but also in "omotenashi" which is the special hospitality found in Japan. Together with isshindoutai, by putting oneself in the shoes of others, as it is no longer "I" or "you" but "we," and by foreseeing possible problem areas so that they can be prevented or at the very least taken care of when they arise, the Japanese have set the norm for making guests and customers feel relaxed and pampered.

It is important to note that the relentless pursuit of perfection or quality of the Japanese comes from the very foundation of the Japanese culture. One of the ways of life of the Japanese can be summed up by what is known as "wabi sabi." The direct translation means "the sadness from the rusting or the passing away of things." The common form of wabi sabi is the cherry blossom viewing in Japan. The Japanese enjoy the petals, which in a very short time begin to fall. The passing of something beautiful that can only be enjoyed for a very short time is a form of wabi sabi. Though it is not common to use this analogy, it may be likened to a candle that burns twice as brightly but lasts for only half the time. This is considered beautiful and sad at the same time which is respected greatly by the Japanese. It is therefore no wonder that the Japanese as a whole love fireworks which are carried out during the summer festivals.

Wabi sabi has its roots in the passing of the seasons. Each passing brings the withering or rusting of nature, sadness, and the start of another season. The haiku, a form of Japanese poetry, has its origins from wabi sabi. Most of the content deals with seasonal foods, activities, and weather. The sadness and the spirit of perfection are clearly seen in what is called "hanseikai." In Western countries, whether it be a piano recital, the completion of a marathon, the success of the business project, etc., there will always be a celebration commemorating the success. With the Japanese, however, there will be a celebration along with criticism on how things could have been done better. No matter if human abilities were surpassed, there would always be a hanseikai. Japanese children are exposed to this from the first grade in elementary school in the form of calligraphy. The children are taught to pursue a better writing style each time they hold the pencil or brush. No matter if one even becomes an expert, there will always be room for more improvement. Those coming from a North American background may find this stressful and almost unbearable as no amount of effort and hard work would be enough. It is never enough and it never ends. This is the sadness or wabi of the Japanese. It's the way of life to embrace this sadness.

Conclusion

It is the combination of all these customs and mind-sets that painted the picture of "Japan quality and efficiency" system. Without wabi sabi, the Japanese would not be able to endure the constant criticism and the never-ending strive for perfection. Without the "group" and the sharing of responsibilities, far less importance would be placed in doing the best that one could do. In addition, without the prevention of possible issues and taking special care of issues at the right time and place along with "wasting not" to keep harmony with nature, an efficient process in manufacturing or other activities would not be possible. However, in order to have more creativity among the Japanese, it would not suffice just to have more lateral movement of activity of the brain through some sort of technique such as lateral eye movement training (Carter 2000; Shobe et al. 2009; Bogen and Bogen 1988), by learning the English language more thoroughly, etc. The Japanese have the tendency to readily accept new technology or ideas from America, but slow to the same things if they originate from their homeland. What is needed is some kind of system or methodology to facilitate creativeness and to have individuality and the freedom to express it without being shunned by the society in Japan.

References

Bogen, J. E., & Bogen, G. M. (1988). Creativity and the corpus callosum. *Journal för Psychiatric Clinics in North America, 11*(3), 293–301.

Carter, R. (2000). *Mapping the mind.* Berkeley, CA: University of California Press.

Dunn, Michael; Greenhill, Simon J.; Levinson, Stephen C.; Gray, Russell D. [2011] Evolved structure of language shows lineage-specific trends in word-order universals Nature Vol. 473 pp. 79–82.

Esaki, M. (2009). *Method for creating wisdom from knowledge.* Tokyo: Japan Material Management Association. Retrieved from http://dtcn-wisdom.jp/Advanced-project-management-methodology.pdf.

Fiebach, Christian J.; Maess, Burkhard; Friederici, Angela D. [2002, last modified 2011, March] Neuromagnetic evidence that differences in noun and verb processing are modulated by the presence of a syntactic context Conference Paper of Max Planck Institute of Cognitive Neuroscience, Leipzig, Germany.

Sereno, Joan A. (1998) Hemispheric differences in grammatical class: A hemifield investigation Working Papers of the Cornell Phonetics Laboratory Vol. 12 pp. 233–242.

Shobe, E. R., Ross, N. M., & Fleck, J. I. (2009). Influence of handedness and bilateral eye movements on creativity. *Brain and Cognition, 71*(3), 204–214.

Tsunoda, T. (1985). *The Japanese brain: Uniqueness and universality.* Tokyo: Taishukan Pub.

Chapter 8
Cross-Cultural Challenges for Innovation Management

Maria Rosaria Della Peruta, Nigel J. Holden, and Manlio Del Giudice

Abstract A conceptual effort presented in this chapter emphasizes the importance of "context" to analyze more in detail the influence of knowledge and communication on the formation of individual and collective interpretive schemes underlying innovative processes. Moreover, from this viewpoint, it is important to understand critical issues associated with the translation of individual and collective cognition into the organizational evolution for a cross-cultural context. As the authors have emphasized in this chapter, creativity, learning, and generating knowledge are premised on cross-cultural collaboration as action.

Introduction

How do managers deal with cross-cultural challenges in the context of innovation management? The traditional response is that they need to create and encourage diversity in organizational thought and action (Del Giudice et al. 2012). It is assumed (e.g., by Tyre and von Hippel 1997) that most of the knowledge that is useful to solve issues or create innovation is tacit and informal and is disseminated through interaction, storytelling, and informal processes in action nets or communal processes. So far, the knowledge management literature does not grasp that the all-important common language arises not just from storytelling or other techniques like brainstorming (Nonaka and Takeuchi 1995), coaching/mentoring (Wenger et al. 2002), conversation (Tarde 1969; Wenger et al. 2002), e-round tables (Mintzberg 2004), dialogues (Nonaka and Takeuchi 1995; Nonaka et al. 2008), metaphorical

M.R.D. Peruta, Ph.D. (✉)
Second University of Naples, Naples, Italy
e-mail: mariarosaria.dellaperuta@unina2.it

N.J. Holden, Ph.D.
Leeds University Business School, Leeds, UK
e-mail: N.Holden@lubs.leeds.ac.uk

M. Del Giudice, Ph.D.
Univeristy of Rome "Link Campus", Via Nomentana, 335-00162 Rome, Italy
e-mail: m.delgiudice@unilink.it

I.N. Dubina, E.G. Carayannis (eds.), *Creativity, Innovation, and Entrepreneurship Across Cultures*, Innovation, Technology, and Knowledge Management, DOI 10.1007/978-1-4939-3261-0_8

discourse (Kohlbacher 2007; see also Zaltman 1997), and organizational languaging (McKenzie and van Winkelen 2004).

From this perspective, learning can be viewed as a process involving individuals in groups on a daily basis. It can be described as the buildup of experience, the creation of skills, and the generation and development of knowledge via practice or action. Scholars who believe in the idea of learning as practice are still convinced that human minds are repositories of knowledge. Nonetheless, they view learning as an inherently social process. But learning, knowledge, and action are not necessarily self-supporting but depend on the context. Knowledge may be expressed in a formal way and thus made explicit. However, and crucially, it is tacitly ingrained above all in individuals' experience and disseminated through direct communication.

Nightingale (2003) uses the notion of "tacit consent" to refer to the way in which tacit knowledge not to mention practical experience is shared. But, that in itself is an insufficient explanation. Here, we need to draw on Searle's (1969) famous and useful concept of "speech acts" in discussion of the interlingual sharing of tacit concepts. It is not just a question of the intelligibility of what we might call shared content of tacit knowledge but the intention that informs the entire context of communication. As the great linguistic scientist, Noam Chomsky (1976, p. 64) has observed: "The theory of speaker's intention may well be a contribution to a theory of successful communication." When countless management tracts aver that effective cross-cultural managers need to have "tolerance to ambiguity," they do not seem to realize that ambiguities arise less from crosslinguistic misunderstandings as popularly understood and more from failures of discernment as to the true nature of intention as reflected in the speech acts of a counterpart who is not just an individual but a person representing the interests of a corporate entity.

This leads to the observation that tacit knowledge is not uniform within and across organizations. Such differences have not been sufficiently stressed by organizational scholars. As Brown and Duguid (2001) have noted, they have often just considered the idea of shared meanings and organizational coherence as given. Yet the capacity to reach shared meanings cannot be understood as a premise but must be considered a consequence of learning activities, which have the purpose of creating an adequate degree of coherence within the organization. This means that shared meanings and organizational coherence are "emergent" systemic phenomena which are neither given nor assumed but need to be explained (Cohen and Sproull 1991; Holland 1992). This, in turn, highlights the importance of thoroughly analyzing the idea of "cross-cultural innovation" in order to examine the latter as an emergent property that needs to be constantly restored. As we have emphasized in this chapter and need to stress again, the learning we are talking about is premised on cross-cultural collaboration *as action*.

Overview of Innovation

To introduce to the understanding of the creative and innovative processes, we must dwell on some basic concepts, namely, invention, innovation, and innovative creativity.

"Perhaps the definition of invention is the most solid definition in the field of creativity" (Huber 1998, p. 59). Invention refers to the birth of an idea for a new product or process. According to Fleming (2007), "at its simplest level, this definition provides an accessible picture of the inventor as a tinker, trying different combinations of materials, gadgets and configurations, and every invention can be thought of as an assemblage of its constituent parts, including the steamship (sailing ship and steam engine), the automobile (bicycle, carriage and internal combustion engine) and Apple Inc.'s iPod (cheap memory, digital music and lightweight battery)."

Innovations that usually include creativity, however, are not the same thing. The innovation may be defined as the successful implementation of creative ideas into business reality. In this view, creativity by individuals and teams is a starting point for innovation; the first is necessary but not sufficient condition for the second (Amabile et al. 1996, 1154–1155)

The term innovation comes from the Latin verb innovate meaning "renew" or "change." So it is clear already from the etymological meaning that innovation does not refer to the discovery of something new. Sahal (1983, p. 215) suggests "the fact that chance plays a central role in discovery. Equally important, the act of creation does not depend upon pure chance alone; the term discovery is used in a neutral sense to include both invention (creation of a new device) and innovation (its commercial application)."

Innovation is the creation and delivery of new value. Creation is an essential part of it—it's got to be new. Delivery is an equally important part of it. It is often difficult to distinguish the three concepts, since they are closely connected. Insights into the patterns of development processes within communities or organizations reflect the different requirements for elaborating ideas and putting them into practice. Basically, inventions can occur anywhere, while innovations are generally carried out by enterprises that "invest considerable amounts of time, money and other resources in the search for new innovative opportunities. Such investment increases the ability to create, use, and recombine new and existing knowledge" (Laursen and Salter 2006, p. 131).

The critical thinking of what are the borders of innovative or inventive or creative processes has been even complicated by the fact that they have been analyzed in an absolute way. Furthermore, it is pivotal to highlight that these processes are continuous. In particular, examining what Joseph Schumpeter said on the emergence of novelty in economic institutions, Becker et al. (2006, pp. 356–357) has essentially focused on his "evolving" vision of economic development: "As he revised the book for subsequent editions in German (1926) and English (1934), Schumpeter apparently changed his views. Gradually, he shifted to a position that de-emphasized the dichotomy between incremental improvement and the introduction of novelty, blending these into a continuum. …In his article *Development* (2005), Schumpeter identified the explanation of novelty as the greatest unmet scientific challenge…. He emphasized the impact of new combinations on competition. Famously, he identified five types of new combinations that mattered in this regard: new products, new production methods, new forms of organization, new markets, and new sources of supply. However, he was never able to link his typology of new combinations to an understanding of the processes generating novelty."

The introduction of a novelty in a particular setting often involves significant adjustment and organizational changes or innovations. "Very soon, the complex conditions of the competitive contests that characterized the last decades of the twentieth century made activation of a latent need for independence and autonomy at the origin of entrepreneurial actions almost unrealistic and proposed new dynamics for the accumulation and sharing of knowledge with a reduction of costs and risks connected to innovation" (Del Giudice et al. 2013, pp. 120–121). These may lead to a strong increase in productivity and competitiveness (see Del Giudice and Straub 2011).

Later literature discussed the matter, although it was initially slow to evolve. Most of the researchers have agreed with Schumpeter's focus on uncertainty (Nelson and Winter 1982; Nonaka and Takeuchi 1995; Van de Ven et al. 1999; Brouwer 2000; Hanusch and Pyka 2007; York and Venkataraman 2010; Anderson and Tushman 2001).

Especially in the case of innovation rewarding, it is difficult to unveil the most significant sources or the best options to pursue or even if the project will be successful (Coombs and Hull 1998; Kogut and Zander 1992; Danneels 2002; Fagerberg 2004). Due to the nature of innovation, it is possible to state that its complexity depends on the scope of all the new combinations of factors, such as ideas, capabilities, skills, and resources, in a given context. "Interactive, collective learning is based on intra- or inter-organizational institutions (routines, norms and conventions) regulating collective action as well as on tacit mechanisms for the absorption of codified knowledge. This requires that the actors in question have tight connections to the 'local codes', on which collective tacit as well as disembodied codified knowledge is based. Thus, depending on the actual architecture of a productive knowledge base, the ability to interpret local codes will be critical for the integration of the operations of a firm within an inter-firm network or production system" (Asheim and Coenen 2006, p. 164). This change can be interpreted as an evolution from production systems to learning systems, and it entails a shift from the traditional view of production systems as fixed flows of goods and services to dynamic systems based on learning.

Hence, the organization for innovation is not a simple task. This way of reinterpreting a business leads to highlighting how the limits to development, to paraphrase Penrose, are no longer represented by growing costs or the lack of opportunity but on the capacity or lack of capacity for managing the fragmentation and distribution of the knowledge necessary for operating successfully in environments continually modified by innovation, competition, and institutional change.

Implications for Cross-Cultural Innovation

It is necessary to take into account the complexity of the innovation process, as an activity that is institutionally constructed and developed. It is important to understand if the potential for communication and interaction through the linkages within the organization is adequately exploited and if other linkages may be created with profit. Certain patterns of interaction and results will be facilitated by the structure of an organization, while others will be constrained; nevertheless, these constraints are not necessarily of a merely technical nature and… "…cannot realistically be reduced to designing flows of codified information across functional boundaries." To realize the full potential of the new innovation, "… involves co-ordinated experimentation…and the interpretation of ambiguous or incomplete data, where tacit knowledge is essential" (Pavitt 1998, p. 14). "This heuristic, subjective and internalized knowledge is not easy to communicate and is better learned through example and practice" (Delmas 1999, p. 640). "It follows that supporting the formation of shared meanings across the organisation is not simply about ensuring smooth information and communication flows across functions, but also, and most importantly, about integrating meaning structures" (D'Adderio 2003, p. 342).

The conceptual effort presented here emphasizes the importance of "context" to analyze more in detail the influence of knowledge and communication on the formation of individual and collective interpretive schemes underlying innovative processes. Moreover, from this viewpoint, it is important to understand critical issues associated with the translation of individual and collective cognition into the organizational evolution for a cross-cultural context.

Our notion of translation is consistent with Czarniawska's (2001) contention that in a given context, the term "translation" has a much wider meaning than that inferable from its literal interpretation. Based on the concept developed by Latour (1993), it can be described as "displacement, drift, invention, mediation, creation of a new link that did not exist before and modifies in part the two agents." In the context of this paper, we may regard for conceptual purposes agents as two cross-culturally interacting parties. Latour (1999) sees such agents taking "detours through the goals of others," whereby both aim to "translate the interest of the other" (p. 89). In such a process, these ad hoc translations give rise to shifts in meaning and perceptions as to objectives and motives. These shifts, the direct result of social interactions, both formal and informal, can in practice be so subtle that they are missed by the other party. Not only do these shifts have a direct effect on the construction of meaning, as jointly created, yet possibly misunderstood by one or even both parties, they become part of the character of what Hutchins (1991, p. 15) calls "distributed cognition," whereby socially disseminated tacit knowing is ideally shared by those who work together and is embedded in their minds and bodies, as well as in artifacts and environmental structures. "While tacit knowledge no doubt plays a small but important role in these processes, these are many more important causal processes at work, such as those relating social interactions between individuals. These cannot be reduced down to aggregated individual tacit knowledge" (Nightingale 2003, p. 178). In these highly common, yet vastly intricate processes, the role of language, on the one hand, and the nature of the actual setting, on the other, influence the ability and willingness to share knowledge.

The literature considering constraints on cross-cultural innovation does not take account of the conviction that its "success" is premised on acts of translation, whereby the creation of a common language is facilitated by cross-cultural collaborative learning, and that common knowledge has context-specific tacit significations and

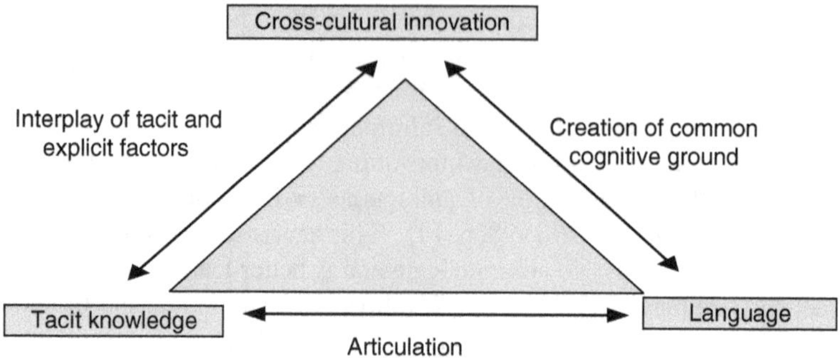

Fig. 8.1 Cross-cultural innovation, tacit knowledge, and language (Holden and Glisby 2010)

knowledge-creating properties. We contend that cross-cultural innovation is best understood as process involving the subliminal influences of tacit knowledge and their relationship with language.

Our task now is to model these otherwise complex processes and relationships in ways that not only usefully condense our convictions and arguments, but also serve as a reliable guide to those in the KM community who seek enlightenment about how to integrate something as elusive as tacit knowledge into their schemes. The model we present here (Fig. 8.1) is a modification of one devised by Holden and Glisby (2010).

It is our conviction that the relationship between tacit knowledge and cross-cultural innovation can *only* be clearly understood if language is brought into the equation. Our golden triangle represents an attempt to capture—in an admittedly simplified form—the interrelationship between the three dynamic factors. Each of the three sides of the triangle reflects key factors associated with the respective relationship. Thus:

- Tacit knowledge cannot become part of any KM system unless it is part of an interplay with explicit knowledge.
- Tacit knowledge cannot be expressed, and ultimately formalized, in language without articulation.
- Language can only be of service to cross-cultural innovation if it contributes to the creation of "common cognitive ground" (Nonaka and Takeuchi 1995).

The model is predicated on the fact that KM in cross-cultural situations, in which protagonists must attempt to create and share knowledge via language barriers of variable permeability, reconcile differences in worldview, and establish common cognitive ground *all at the same time*, is confronted with significant—and often unanticipated—challenges at the personal and organizational levels of interaction. Not only that, pragmatically speaking, the communicative behavior of actors whatever their national, ethnic, or professional background, is "spontaneous," ad hoc, and improvised (Holden 2014). The model also recognizes that tacit knowledge is an integral element of cross-cultural innovation and associated processes, all of which involve the continuous use of language.

The model makes it clear that there is direct interdependency between knowledge management, on the one hand, and language and tacit knowledge, on the other. As a conceptual contribution to cross-cultural knowledge management, this is important for two reasons. First, there is no consistent view as to the role of language in the KM literature, while the issue of relative differences among languages as modes of expression and stores of tacit knowledge is still not much discussed. Second, disputing the numerous references to tacit knowledge in the KM literature, discussion is often cursory. In support of that proposition, Holden and Glisby (2010) surveyed 12 books on KM, which totaled 3000 pages of text. One edited book about KM best practices in Europe had no references. In total, there were a mere 156 references to tacit knowledge. In the case of two books, Nonaka was the lead author of one of these books and one of three joint authors of the other. These two books accounted for 81 of the indexed references or 52 % of the total.

Spender (2013) credits Nonaka with creating "completely new language" and attributing to middle management "a newly identified middle-up-down coordinating and language management role." His major work, *The knowledge-creating company*, "created new managerial language" … "added further new language [i.e. terms]—specifically the terms *ba*, *flow*, and *phronesi*" (p. 45). In a similar vein, Campbell (2013) has argued that Nonaka's "main contribution is a language for describing capabilities in divisionalised organizations and a typology of different roles corporate headquarters any play in building core capabilities." In his commentary on Nonaka's contribution, Kohlbacher (2013, p. 296) has stressed "the importance of language for knowledge creation as well as the critical issue of re-creating and refining tacit knowledge over time."

Our model does not merely attempt to represent the interdependencies, but indicates how conversion from one state to another (i.e., language to knowledge management, tacit knowledge to knowledge management, and tacit knowledge to language) requires activities. As with all such representations, the model is a very considerable simplification of actual processes. It does, however, aspire to be more than a notional model. It can be mapped onto *any* cross-cultural exchange of knowledge at the individual level of interaction as well as the corporate one. With respect to cross-cultural innovation, the model has the benefit of drawing attention away from acts of sharing "new" knowledge to subliminal influences which are variously embedded in language, contexts, and people's minds and underpin, often beyond management control, the entire innovation process.

From Cross-Cultural Innovation to Cross-Cultural Management

As highlighted by Giddens (1979), there are at least two levels of consciousness in every social system: discursive and practical. The first is related to knowledge formulated by actors discursively. The second implies tacit stocks of knowledge that cannot be usually expressed by actors at the level of discourse, but are based on their social behavior. Due to the fact that meaning systems generally function beyond the limits of consciousness, it is often automatically thought that the consonance between one's opinions and those of others is higher than what it is found in reality. This process of reciprocal perspective taking is quite elaborate because of the tacit nature of at least some of the beliefs, knowledge, norms, and meaning systems that establish structurational or interpretative conventions. Thus, reciprocal comprehension implies something that cannot be normally explained in an explicit way. People are not usually aware of such content in their daily lives because it is contained and embedded in structurational conventions.

Particular "constellations of meaning" are the result of language itself; within those constellations, the world is conceptualized by people, facts are organized, and knowledge is generated. Thus, through language and meaning, human thought is altered and action guided. The abovementioned constellations of language and meaning underlie facts and knowledge and intensely affect wider social institutions and cultural practices.

From a cross-cultural viewpoint, these assumptions hinder the process of effectively acquiring each other's perspective, making learning difficult. The basic issue is that people coming from different cultural backgrounds will probably interpret information on the basis of their specific systems of meaning and their diverse stocks of knowledge. It is therefore inevitable to assume that in multiethnic firms, tacit knowledge may be transferred with difficulty, and it may be even more complex to synthetize creatively different knowledge bases.

In the process of intercultural exchange, the specific beliefs, values, knowledge, meanings, and assumptions of an individual are transferred, assessed, and integrated with the corresponding beliefs, values, knowledge, meanings, and assumptions of other individuals. Thus, it becomes fundamental to guarantee some social balance within organizations, so meaning systems can emerge and everyone can easily understand each other; to do so, a set of rules must be implemented.

As part of this frame, it is possible to usefully interpret the specificity of ethnic enterprises. Multiethnic people have been taught by the ethnic entrepreneurship experience that it is preferable to discard specific customs of the different places of origin in order to become part of a much wider affiliation. As pointed out by Mantzaris (2000), this affiliation is obtained through the use of a common language that allows mutual communication. Therefore, as stressed by Koliopoulos and Veremis (2002), the idea at the basis of the creation of a wide united group of immigrants is no longer the place of origin but mainly the use of a common language.

Ethnic enterprises are economic institutions that often exist prior to the economic activity because they are based on assumptions, ties of value, culture, honor, trust, and respect, exceeding the merely economic significance of social aggregation. In general terms, to explain differences in entrepreneurial operations between ethnic minority groups, two fundamental variables are taken into account (Waldinger et al. 1990, 2000). The first regards cultural (pull) factors, which are related to ethnicity and in particular collective ethnic properties. This perspective concerns the way entrepreneurship is viewed in diverse cultures and specifically how much people are inclined to create an enterprise in their country of origin. In terms of cultural explanations, the indigenous characteristics of the ethnic entrepreneur are introduced as a way of stressing differences between ethnic groups in terms of business activity.

The structural (push) view, instead, analyzes the differences in business activities in structural factors on the labor market. For example, immigrants are often subject to discrimination on the labor market. The consequence is that structural factors correspond to the meaning of contextual opportunity structures. This argument entails that the significance of ethnicity changes according to the context. However, everyone has an indisputable ethnic identity, depending either on their well-identified ethnic group (Arabs, Jews, for instance) or on their specific country of origin (e.g., Iran, Israel). All multiethnic companies hire people with different objective ethnic backgrounds. Though, there is no necessary correspondence between objective ethnic identity or background and the subjective identity that appears in business operations of immigrant entrepreneurs.

Therefore, if one is willing to comprehend what objective ethnic identity or background means for multiethnic companies, it is fundamental to analyze how ethnicity is understood in the empirical context, since a simple predefinition is not sufficient. This can be interpreted as a performative perspective which addresses practice rather than theory. In summary, as stated by Hosking (1999), a performative approach requires that the actors and the context determine the way to understand and give meaning to the subject they are dealing with. The way to achieve such a comprehension is by using language to communicate (Burr 1995) because it is in this way that meaning is built (Fletcher 2003). The focus is on social interaction, not individual choices. This implies participation in what occurs in a specific context in which the preferences of a single actor are not the only ones involved. In fact, overall performance has the same significance as what is perceived by the single actors in relation to what they do. When business is organized by entrepreneurs according to the interpretation of particular cultural values and beliefs, personal experience and actions may be severely evaluated together with the experience and actions of those with different cultural backgrounds.

Rather than searching for the best sources of knowledge possible, people are generally prone to trust those who appear and behave like them. Bartolomeu whose company imports food products from Eastern Europe and supplies shops and restaurants in Italy is an entrepreneur with a large employee demographic from both Somalia and Romania. He worked with employees that were not "from here." They had no family and no cultural connections in whom to find refuge and a sense of belonging.

As stated by Kirsh (1999), the multiethnic enterprise can be helpful in explaining the coexistence in the same person and in the same community of very different values, but it is also important to comprehend workplace dynamics as an elaborate supposition of social, cultural, cognitive, and physical constraints. From the perspective of distributed cognition (Hollan et al. 2000; Hutchins 1991), people who work together share socially distributed tacit knowing, which is embedded in artifacts and environmental structures and embodied in their minds and bodies. As a consequence of social interaction in a setting in which internal and external structures are involved, the capacity and propensity to share knowledge are driven by culture and become context sensitive.

However, it is certainly important to bear in mind that individuals are more complex and unpredictable than cultural norms. Knowledge of a norm does not necessarily imply an inevitable effect, although it does entail its recognition and verification when this occurs. Nevertheless, it is very complicated to change cultural norms, and when a situation of threat seems to arise, individuals will generally return to the cultural expectations they are accustomed with.

When one realizes the value trump possibility, one could still not agree with the value choice; however, the cultural logic underlying it will be understood better. Communication can be adjusted according to the new cultural expectations so there is mutual comprehension.

Thus, cultural adaptation is influenced by one's own attitude, the attitude of the host culture and that of the ethnic group of origin, together with the capacity to learn and grow or even change. In the workplace, it is important not only to adapt to the country culture and to the cultures of the people one works with, but also to the business or organizational culture. This is why, when approaching a new workplace, it is essential to discover new ways to learn about the new environment, without removing one's own beliefs. Individuals who manage to adapt better to the new context experience positive outcomes: they are happy about their own cultural group and also about the new culture they are becoming part of. This is both the result of the surrounding context and of personal attitude and behavior.

Conclusion

It is possible to argue that the cultural factor functions as a rationalized institutional myth, which is a value generally considered in a positive light, and therefore reinforced and structurally embedded in firms with its principles and rules, since it is deemed legitimate from the outset, without evaluating its influence on the organization's performance.

Thus, in this view, companies that acquire cross-national features over time tend to behave isomorphically and consolidate the set of values and norms recognized by their members, since they are constrained within an increasingly firm framework of cultural perceptions. One important consequence of these observations that a concept of cross-cultural management that sees the challenge in terms of coping with different value systems is wholly inadequate for the kind of dynamic and ever-changing character of cross-cultural innovation behavior, which we have described. A starting point for a more perceptive notion is based on Bartholomew and Adler's (1996) conviction that the keys to the kingdom are to be found in "cross-cultural collaborative learning." But very little will in fact be learned unless individuals and organizations ascribe greater power to tacit knowledge as a directly shaping influence on the quality of interactions and therefore on the outcomes of *any* cross-cultural collaborative behavior.

References

Amabile, T. M., R. Conti, H. Coon, J. Lazenby, and M. Herron. (1996). Assessing the Work Environment for Creativity. *Academy of Management Journal, 39*(5), 1154–1184.

Anderson, P., & Tushman, M. L. (2001). Organizational environments and industry exit: The effects of uncertainty, munificence and complexity. *Industrial and Corporate Change, 10*(3), 675–711.

Asheim, B. T., & Coenen, L. (2006). Contextualising regional innovation systems in a globalising learning economy: On knowledge bases and institutional frameworks. *The Journal of Technology Transfer, 31*(1), 163–173.

Bartholomew, S., & Adler, N. (1996). Building networks and crossing borders: The dynamics of knowledge generation in a transnational world. In P. Joynt & M. Warner (Eds.), *Managing across cultures: Issues and perspectives* (pp. 7–32). London: International Thomson.

Becker, M. C., Knudsen, T., & March, J. G. (2006). Schumpeter, winter, and the sources of novelty. *Industrial and Corporate Change, 15*(2), 353–371.

Brouwer, M. (2000). Entrepreneurship and uncertainty: innovation and competition among the many. *Small Business Economics, 15*(2), 149–160.

Brown, J. S., & Duguid, P. (2001). Knowledge and organization: a social-practice perspective. *Organization Science, 12*(2), 198–213.

Burr, V. (1995). *An introduction to social constructionism*. London: Routledge.

Campbell, A. (2013). Building core capabilities: the roles of corporate headquarters. In G. Von Krogh, K. Kase, & C. Cantón (Eds.), *Towards organizational knowledge: The pioneering work of Ikujiro Nonaka* (pp. 221–240). London: Palgrave-Macmillan.

Chomsky, N. (1976). *Reflections on language*. Glasgow: Fontana/Collins.

Cohen, M., & Sproull, L. (1991). Special issue on organizational learning. *Organization Science, 2*(1), 1–147.

Coombs, R., & Hull, R. (1998). Knowledge management practices' and path-dependency in innovation. *Research Policy, 27*(3), 237–253.

Czarniawska, B. (2001). Is it possible to be a constructionist consultant? *Management Learning, 32*(2), 253–266.

D'Adderio, L. (2003). Configuring software, reconfiguring memories: the influence of integrated systems on the reproduction of knowledge and routines. *Industrial and Corporate Change, 12*(2), 321–350.

Danneels, E. (2002). The dynamics of product innovation and firm competences. *Strategic Management Journal, 23*(12), 1095–1121.

Del Giudice, M., Carayannis, E. G., & Della Peruta, M. R. (2012). Culture and cooperative strategies: Knowledge management perspectives. In *Cross-cultural knowledge management* (pp. 49–62). New York: Springer.

Del Giudice, M., & Straub, D. (2011). IT and entrepreneurism: an on-again, off-again love affair or a marriage? *MIS Quarterly, 35*(4), 3–8.

Delmas, M. A. (1999). Exposing strategic assets to create new competencies: the case of technological acquisition in the waste management industry in Europe and North America. *Industrial and Corporate Change, 8*(4), 635–671.

Fagerberg, J. (2004). Innovation: a guide to the literature. In J. Fagerberg, D. C. Mowery, & R. R. Nelson (Eds.), *Oxford handbook of innovation*. Oxford: Oxford University Press.

Fleming, L. (2007). Breakthroughs and the "long tail" of innovation. MIT Sloan Management Review 49 (1): 69–74.

Fletcher, D. (2003). Framing organizational emergence: discourse, identity and relationship. In C. Steyaert & D. Hjorth (Eds.), *New movements in entrepreneurship* (pp. 125–142). Cheltenham: Edward Elgar.

Giddens, A. (1979). *Central problems in social theory: Action, structure and contradiction in social analysis*. London: The Macmillan Press Ltd.

Hanusch, H., & Pyka, A. (2007). Principles of neo-Schumpeterian economics. *Cambridge Journal of Economics, 31*(2), 275–289.

Holden, N. (2014). The Consequences of Culture in International Business: A Long Overdue Commentary. Journal of Business Anthropology, 3(2), 180–190.

Holden, N. J., & Glisby, M. (2010). *Creating knowledge advantage: The tacit dimensions of international competition and cooperation.* Copenhagen: Copenhagen Business School Press.

Hollan, J., Hutchins, E., & Kirsh, D. (2000). Distributed cognition: toward a new foundation for human-computer interaction research. *ACM Transactions on Computer-Human Interaction (TOCHI), 7*(2), 174–196.

Holland, J. (1992). *Adaptation in natural and artificial systems.* Cambridge, MA: MIT Press.

Hosking, D. M. (1999). Social construction as process: some new possibilities for research and development. *Concepts and Transformation, 4*(2), 117–132.

Huber, J. C. (1998). Invention and inventivity as a special kind of creativity, with implications for general creativity. *The Journal of Creative Behavior, 32*(1), 58–72.

Hutchins, E. (1991). Organizing work by adaptation. *Organization Science, 2*(1), 14–39.

Kogut, B., & Zander, U. (1992). Knowledge of the firm, combinative capabilities, and the replication of technology. *Organization Science, 3*(3), 383–397.

Kohlbacher, F. (2007). International marketing in the network economy: a knowledge-based approach. Palgrave Macmillan.

Kohlbacher, G. (2013). Knowledge-based marketing: fostering innovation through marketing knowledge creation. In G. Von Krogh, K. Kase, & C. Cantón (Eds.), *Towards organizational knowledge: The pioneering work of Ikujiro Nonaka* (pp. 283–305). London: Palgrave-Macmillan.

Koliopoulos, J., & Veremis, T. (2002). *Greece: The modern sequel.* London: Hurst.

Kirsh, B. (1999). Organizational culture, climate and person-environment fit: Relationships with employment outcomes for mental health consumers. Work (Reading, Mass.), 14(2), 109–122.

Latour, B. (1993). *On technical mediation: The messenger lectures on the evolution of civilization.* Lund: Lund University.

Latour, B. (1999). *Pandora's hope: Essays on the reality of science studies.* Cambridge, MA: Harvard University Press.

Laursen, K., & Salter, A. (2006). Open for innovation: the role of openness in explaining innovation performance among UK manufacturing firm. *Strategic Management Journal, 27*(2), 131–150.

Mantzaris, E. A. (2000). *A social history of the Greek community in South Africa.* Unpublished manuscript.

McKenzie, J., & Van Winkelen, C. (2004). Understanding the knowledgeable organization: nurturing knowledge competence. Cengage Learning EMEA.

Mintzberg, H. (2004). Managers, not MBAs: A hard look at the soft practice of managing and management development. Berrett-Koehler Publishers.

Nelson, R. R., & Winter, S. G. (1982). *An evolutionary theory of economic change.* Cambridge, MA: Harvard University Press.

Nightingale, P. (2003). If Nelson and Winter are only half right about tacit knowledge, which half? A Searlean critique of "codification". *Industrial and Corporate Change, 12*(2), 149–183.

Nonaka, I., & Takeuchi, H. (1995). *The knowledge-creating company.* New York: Oxford University Press.

Nonaka, I., Toyama, R., & Hirata, T. (2008). Managing flow: A process theory of the knowledge-based firm (Vol. 19). New York: Palgrave Macmillan.

Pavitt, K. (1998). Technologies, products and organization in the innovating firm: what Adam Smith tells us and Joseph Schumpeter doesn't. *Industrial and Corporate change, 7(3),* 433–452.

Sahal, D. (1983). Invention, innovation, and economic evolution. *Technological Forecasting and Social Change, 23*(3), 213–235.

Schumpeter, J. A. (1926). Theorie der wirtschaftlichen Entwicklung: Eine Untersuchung über Unternehmergewinn, Kapital, Kredit, Zins und den Konjunkturzyklus.

Schumpeter, J. A. (1934). *The theory of economic development. An inquiry into profits, capital, credit, interest, and the business cycle*. Cambridge, MA: Harvard University Press.

Schumpeter, J. A. (2005). Development. *Journal of Economic Literature, XLIII*, 104–116.

Searle, J. R. (1969). *Speech acts: An essay in the philosophy of language*. Cambridge: Cambridge University Press.

Spender, J. C. (2013). Nonaka and KM's past, present and future. In G. Von Krogh, K. Kase, & C. Cantón (Eds.), *Towards organizational knowledge: The pioneering work of Ikujiro Nonaka* (pp. 24–59). London: Palgrave-Macmillan.

Tarde, G. (1969). The public and the crowd. On Communication and Social Influence. Selected Papers edited and with an Introduction by Terry N. Clark, Chicago, 277–294.

Tyre, M. J., & Von Hippel, E. (1997). The situated nature of adaptive learning in organizations. *Organization Science, 8*(1), 71–83.

Van de Ven, A.H., Polley, D.E., Garud, R., Venkataraman, S. (1999). The Innovation Journey. Oxford University Press, New York.

Waldinger, R., Aldrich, H., & Ward, R. (1990). *Ethnic entrepreneurs: Immigrants and ethnic business in western industrial societies*. Thousand Oaks, CA: Sage Publications.

Waldinger, R., Aldrich, H., & Ward, R. (2000). Ethnic entrepreneurs. In R. Swedberg (Ed.), *Entrepreneurship. The social science view*. Oxford: Oxford University Press.

Wenger, E., McDermott, R. A., & Snyder, W. (2002). Cultivating communities of practice: A guide to managing knowledge. Harvard Business Press.

York, J. G., & Venkataraman, S. (2010). The entrepreneur – environment nexus: uncertainty, innovation, and allocation. *Journal of Business Venturing, 25*(5), 449–463.

Zaltman, G. (1996). Metaphorically speaking. *Marketing Research, 8*(2), 13–20.

Chapter 9
Cultural Influences on Innovation and Competitiveness

Panagiotis E. Petrakis

Abstract Innovation is not just a result of R&D activities, the use of advanced technology, and science-based industries. Above all, it is a vehicle for the creation of future entrepreneurship advantage. A key way to realize an environment conducive to continuous innovation is by creating a cultural environment in which innovation can continuously flourish. Creating an innovation culture offers an important source of competitive advantage, both for individual enterprises and for the economy as a whole. This chapter shows that the cultural background of societies can boost economic innovation and thus accelerate competitiveness.

Culture, Social Capital, and Knowledge

The term "social capital" reflects the cultural characteristics (Fukuyama 1999) inherent in the economic and cultural institutions of societies, as well as their sense of justice and rationality. Moreover, increased civic participation correlates positively with efficiency in the functioning of institutions (Putman 1993). Degree of loyalty, as an essential component of social capital, impacts financial results (Arrow 1972; Fukuyama 1995).

Social capital as property, whether of individuals or groups, is associated with many important effects. From the perspective of the individual, social capital can be considered a remarkable set of secondary resources that increase the power of the individual and make it easier to achieve their goals. These can include not only economic goals but also personal goals and related milestones (Snijders 1999).

Particularly interesting to most researchers are the collective effects of social capital. They include, inter alia, the positive effects on public administration at all levels, as well as on national economic performance, the smooth and efficient functioning of modern economies and other broader social issues (Halpern 2005). These effects can generally be traced to the impact on social infrastructure and especially

P.E. Petrakis, Ph.D. (✉)
Department of Economics, University of Athens, Greece,
Stadiou S Str., PC 10562 Athens, Greece
e-mail: ppetrak@econ.uoa.gr

I.N. Dubina, E.G. Carayannis (eds.), *Creativity, Innovation, and Entrepreneurship Across Cultures*, Innovation, Technology, and Knowledge Management, DOI 10.1007/978-1-4939-3261-0_9

to the networks, norms, and trust that define social capital. Social capital thus can be considered a major contributor to growth.

At the political level, social capital contributes to the smooth functioning of the free market and democracy. A lack of social capital leads to disorder in political operations. Civic norms can improve financial results through political channels (Knack and Keefer 1997). High social capital stock is a prerequisite for a civil society, which in turn is a prerequisite for a modern democracy (Gellner 1994). A civil society tends to mitigate the power of the state and protect citizens from that power. Simultaneously, social capital reflects the cultural elements of modern societies (Fukuyama 1999).

The cultural background of societies, and more specifically cultural differences, creates two types of reactions to the impact of cultural differences on how enterprises operate:

(a) The first type of reactions concerns the traditional approach, whereby cultural background becomes a barrier between groups of people with different cultural traits, creating an incentive for eliminating cultural differences.
(b) The second type of reactions perceives cultural background as a source of knowledge and interaction between individuals that can affect the way enterprises operate.

Under this second approach, cultural background can be considered a resource for knowledge management (Holden 2010) and provide a basic instrument that managers can use rather than an obstacle.

Organizational culture is a key factor in the diffusion of knowledge at the firm level (Davenport et al. 1998; DeLong and Fahey 2000), since it can either encourage or impede effective knowledge management. Existing cultural values, beliefs, and stereotypes influence the formation of the social context, which in turn allows the incorporation of those beliefs into knowledge management behaviors. Therefore, the dominant organizational culture can influence knowledge management, while also creating appropriate incentives for team members to produce and spread knowledge. Meanwhile, knowledge management can form part of the organizational culture as it reflects organizational values. The organizational changes that encourage knowledge management to implement strategic changes (such as the introduction of new technology) are affected by the existence of cultural values that can either contribute to this change or make its implementation difficult.

Simultaneously, the identification and utilization of entrepreneurial opportunities are important to the entrepreneurial process (Petrakis and Kostis 2012). However, regardless of the methodology used to identify entrepreneurial opportunities, two main factors initially seem to play a role in identifying entrepreneurial opportunities and activating them as entrepreneurship events: knowledge and cultural background of societies.

Evolutionary Theory and R&D

Veblen (1898) is considered the father of evolutionary economics and noted that "evolutionary economics must be the theory of a process of cultural growth as determined by the economic interest, a theory of cumulative sequence of economic institutions stated in terms of the process itself." Drawing from Darwin and others the imperative that the causal origin of all evolved phenomena had to be explained, Veblen considers future social and economic development the result of collective change in society and institutions rather than change at the individual level. He sees the process of growth and development as a "process of cultural growth," thus making clear that this development does not necessarily represent a qualitative improvement. Decisions taken by individuals can lead to an evolutionary process, but the fairness of the results of their actions is a matter of opinion.

Focusing on the expense of R&D was the basis for restructuring the economic theory through a more evolutionary approach, which contributes to decision making by business executives under uncertainty conditions. The terms "technology," "organization," and "change" are very important in the context of management and the theory of evolution and provide business executives with a different, more interesting, and useful way of economic thinking.

Evolutionary theory involves basic principles, which are also useful for business executives seeking to more efficiently making effective decisions regarding the development of their entrepreneurial plans. Initially, it is reasonable to use an index that links R&D expenses to total enterprise sales, thus showing the intensity and effectiveness of R&D spending. Moreover, in the context of business and strategic development, R&D plays an essential role, focusing on expenses directed to knowledge accumulation and the development of know-how, factors that increase the opportunity for an enterprise to not only survive but grow when facing intense competition (Nelson and Winter 1982; Helfat 1994). According to evolutionary theory, enterprise decisions on R&D spending are shaped by previous decisions and their results. Finally, an accumulation of significant and persistent cross-industry differences in the intensity of R&D spending is observed, while the existence of underlying pressures from the economic and technological environment plays an important role in the emergence of these differences.

Finally, from the perspective of management theory, evolutionary theory focuses on enterprises and the problems they face when operating in a competitive environment. Evolutionary theory does not just accept but urges extensive research involving corporate processes and operations, focused on company organization, the type and quantity of its products and services, the manner it uses to allocate and evaluate expenses, as well as promotional prices and the appropriate direction of movement for R&D expenses to create a development model. It also leads to the detection of proper development directions, especially if concern exists about whether the company has the capabilities to successfully operate in the open economy. Finally, it considers that entrepreneurs and business executives with great skills and experiences and open to challenges may prove very useful in decision making under uncertainty.

Culture, Innovation, and Competitiveness

Differences between countries in levels of technological development and investment capacity are associated with the dominant cultural background of society (Petrakis 2014). The grid of values in a society plays a special role in motivating and encouraging its members to undertake innovative initiatives. This grid refers mainly to levels of individuality/collectivity, as well as the extent of risk aversion, future orientation and planning, acceptance of inequalities, the importance attributed to science in the educational system, and positive attitudes toward science. Simultaneously, given innovation is a complex process that involves learning and the development of new ideas, culture is undoubtedly integral to innovation. Societies, and by extension enterprises and organizations, that wish to be competitive should be innovative (Hana 2013).

In societies characterized by aversion to uncertainty—unlike those with a tendency to uncertainty—there is no long-term loyalty, and so an aversion to innovative activities is observed, with this aversion itself being a long-term investment activity.

Additionally, the ability of enterprises to cope with different cultural aspects to achieve better results is a critical issue. Culture can encourage innovation and creativity. It motivates the individual or group to be entrepreneurial. Different cultural characteristics help the group adopt new perspectives, unlike in groups comprising individuals with similar characteristics.

Obstacles to technological progress, besides cultural traits, can also be social factors such as the social structure of the group, the type of family, interpersonal relations, motivation to team members, and the learning process (Foster 1962). Cultures that reward creativity and encourage their members to achieve personal progress tend to achieve better innovation results. Furthermore, the degree of innovation relates directly to the degree of acceptance of entrepreneurship in society.

Willingness to assume entrepreneurial risk, long-term planning, and acceptance of change are key cultural traits that boost innovation (Rothwell and Wissema 1986).

Furthermore, different behaviors have been observed in relation to business formation, depending on cultural background (Shapero and Sokol 1982). A positive correlation exists between individuality and innovation potential, since the freer individuals are to express their opinions, the more opportunities they enjoy to explore new ideas (Barnett 1953). Individualistic societies tend to encourage their members to express their opinions, and such freedom is necessary for creativity. Furthermore, weak cohesion bonds between members of individualistic societies promote easier dissemination of information than in collectivist societies whose members utilize information at the intergroup level and impede its dissemination. Characteristics of individual mental condition, such as independence, achievement, and encouragement of innovation, are commonly found in individualistic societies (Shane 1992).

Weak power distance and strong individualism also appear related to innovation. Societies with high scores on the dimension of individualism and low scores on power distance achieve higher growth rates and display a stronger tendency for innovation (Hofstede 1984). Societies with high power distance have more centralized governance structures than do low-power distance societies in which decentralized hierarchical structures are preferred, and the latter climate contributes to innovative activity.

Simultaneously, an effective way to maintain the future competitive advantage of enterprises and organizations is through the development of an entrepreneurial culture that boosts innovative activity and thus maintains competitive advantage. Culture should be considered a strategic instrument for achieving competitive advantage through innovation (Petrakis et al. 2014). Innovation cannot simply occur but requires a broader enabling environment (i.e., a culture) that promotes the free exchange of ideas among those engaged in an enterprise—from customers to senior executives. Creating an innovation culture can be an important source of competitive advantage for an enterprise. Culture can encourage innovation and creativity and motivates the individual or group to foster the spirit of entrepreneurship. Different cultural characteristics help the group adopt new perspectives, as opposed to groups composed of individuals who share similar characteristics.

An efficient innovation culture depends on the ability to identify key factors that drive innovation. One way to foster such an innovation culture is through appropriate leaders who clarify targets and how to seek desired outputs to all participants. Focusing on results releases important energy that enables creativity.

The innovation process and gaining of competitive advantage is a complex procedure whose effectiveness is not limited to material resources but has deeper origins. Cultural background is a key factor for determining competitiveness and innovation via two channels: social learning and organizational structure. These channels sometimes prevent and sometimes enhance the diffusion of innovation and competitiveness.

The first channel, social learning (Bandura 1963, 1971), investigates how environmental and cognitive factors interact and affect human learning, preferences, and behavior. Social learning is a dynamic process that transmits cultural characteristics from one generation to the next. Simultaneously, the existence of social stereotypes in every society creates conditions that affect (positively or negatively) the diffusion of knowledge and technology. The prevailing cultural background is of particular interest, since it creates components that either promote or prevent social learning. In other words, the prevailing cultural background forms a grid of interacting factors that ultimately specify human behavior and preferences.

Social learning theory focuses on modeling human behavior through the observation of attitudes and the emotional reaction of individuals. This theory becomes particularly important as it deals with how environmental and cognitive factors interact and influence human learning, preferences, and behavior. It also focuses on the learning that occurs in social frameworks considering that people learn from each other via observation, imitation, and modeling. Furthermore, learning results from the observation of rewards and punishment—known as vicarious reinforcement. The theory extends to traditional behavioral theories, where behavior is governed exclusively by reinforcement, emphasizing the importance of internal processes in the learning process.

We can integrate the evolution of innovations in a social context (Pohlmann 2005), since this process depends on human factors inside or outside the social system. Various factors exert an influence in this direction. The legacy of history helps the individual to distinguish the new from the old, while knowledge and information are spread throughout the community by innovation. Since social learning incorporates the knowledge and experience that people obtain through social relationships and communication, it in turn affects human behavior and decision making.

Cultural background cannot be separated from social learning. Social learning is the medium through which the special characteristics of the society are transferred from generation to generation. It includes tradition, beliefs, and social values and is the modulator of human behavior. Therefore, the focus should be on the contribution of social learning to the decision-making process and economic theory.

Focusing on the cultural background of society, we can identify the forces involved in the process of decision making as it affects individual ability to understand (cognition) and thus the processing of available information. Therefore, it is crucial to identify constructive forces in the evolution of society that promote progress, as well as forces that represent obstacles.

The dimension of individuality/collectivity discussed by Hofstede (1980, 1991) describes the degree to which people learn to act more as individuals than members of a group. Conversely, collectivist societies emphasize the development of cohesion among members. In such societies, individuals possess a team spirit and prioritize the interests of the team over those of individuals. The members of such societies are interdependent, as opposed to the situation in individualistic societies where the bonds between individuals are loose and people are simply concerned with their personal interest. Individualistic societies attach great importance to individual achievement and freedom. In line with Hofstede's individualism/collectivism, the in-group collectivism by House et al. (2004) reflects the degree to which members of a society express loyalty and belief in the institution of family or in the organization to which they belong.

The relationship between individualism and social learning is important for interpreting human behavior. We understand individualism as the feature that constrains the transmission of social learning. Focusing on the dimension of individuality/collectivity, we can extract important aspects from the manner in which social learning is communicated.

The degree to which a society is individualistic is expected to affect the effectiveness of knowledge transmission. Individualistic societies create barriers to knowledge diffusion, since their members are very distant and do not easily enter into social relations. In such cultures, limited social bonds act as a barrier to the knowledge diffusion. Personal interest also prevents extroverted behavior.

House et al.'s (2004) performance orientation is an important dimension that shapes the cultural background of every society. Performance orientation denotes the degree to which a society encourages its members to improve their performance. The performance orientation of the members of the society creates incentives for the development of behaviors that focus on individual interests by limiting knowledge diffusion.

Special reference should also be made to the degree of interpersonal trust[1] and trust in institutions. Besides the fundamental values shaping human behavior, the contribution of trust in every society also possesses crucial social aspects (Almond and Verba 1965; Tocqueville 1990). The existence of social trust creates cohesion bonds among members of a society, and these in turn increase the efficiency of social learning.

[1] The World Values Survey (in all its waves 1–6) measures interpersonal trust using the percentage of people who positively answered the question "Generally speaking, would you say that most people can be trusted or that you cannot be too careful when dealing with people?" The greater the number of people answering that most people can be trusted, the higher the level of generalized trust in the society.

Organizational structure, the second channel, is inextricably linked to enterprise culture, since it facilitates more efficient governance structures and favors the concentration of innovation activity and the gaining of competitive advantage. It is no coincidence that enterprises with large numbers of employees are pioneers in R&D (Kamien and Schwartz 1982; Baldwin and Scott 1987; Cohen and Levin 1989; Cohen 1995).

Organizational structure and culture are among the most important factors in understanding human behavior within organizations and enterprises. Understanding the relationship between these two variables is particularly useful as it helps to determine the behavior of organization members.

Druker (1968) describes structure as the "means for attaining objectives and goals of the organisation." Handy (1993) states that structure is "the allocation of formal responsibilities, the typical organisation chart. It also covers the linking mechanisms between the roles, the coordinating structures of the organization."

Organizational culture determines the way in which team members understand and meet the conditions of the external environment, as well as the way they behave. Hofstede (1980) determines organizational culture as the "collective programming of the mind, which distinguishes the members of one category of people from another." The existence of an organizational culture drives team members to find solutions to specific problems, adapt to the environment, and work together through shared cultural values (Blackwell 2006).

The importance of organizational culture is easily understood, since it contributes to the efficient functioning of activities such as productivity, innovation, and financial performance. Meanwhile, culture can influence employee motivation, goodwill, quality of work, creativity, and workplace attitude (Campbell et al. 2002).

Burns and Stalker (1961) indicate that different types of organizational structure may be effective under different conditions. Additionally, they argue that since the innovation process moves on from the generation of ideas to their implementation, the organizational structure should be differentiated. However, the resistance of organizational structure to change affects the innovation process.

This leads to the question of how innovation is developed and effectively applied within an organization. Answering this question requires searching for conditions that shape the requirements and incentives for developing innovation. We merely have to look for such requirements in the organizational culture. Given that each organizational culture is unique, its effectiveness in creating and exploiting innovation varies.

Initially, an enterprise must innovate to be able to enter the market and be competitive. However, as the organization gains market share, its administrative team becomes increasingly cautious. Many firms settle for a culture of productivity and efficiency and allow innovation to completely disappear.

The main reason an innovation culture is required is that innovation cannot just happen but requires a broader enabling environment that promotes the free exchange of ideas among those engaged in an enterprise—from customers to senior executives. To achieve this, managers should create procedures for the collection, evaluation, and funding of innovative ideas that will create the conditions for an innovation culture.

Even where appropriate ideas exist in the context of an enterprise's operations (e.g., ideas for streamlining internal processes, improving existing product lines, and enhancing connections to customers or maybe even for breakthrough technologies or entirely new products or services), maximizing the number of opportunities created requires the development of a system that seeks, captures, assesses, classifies, and then focuses on and implements the most promising opportunities. Simultaneously, a mechanism is needed for the dissemination of ideas within the organization that allows for the diffusion of ideas, creative thinking, and, ultimately, innovation.

The Great Recession in Europe as a Case Study

In economies most affected by the crisis, a reduction of innovation capacity and thus a deterioration of competitiveness are expected. However, this conclusion is not supported by empirical testing, as conducted by Petrakis et al. (2014) using classification of data, trend analysis, and ordinary least squares. Their study used the recent Great Recession (2008–2013) as a case study of severe deterioration in macro-conditions. On the one hand, they found that countries that were lagging (progressing) with respect to innovation and competitiveness in 2008 continued to lag (progress) to the same or a greater degree 5 years later. Moreover, this effect persisted regardless of macroeconomic changes that accompanied the Great Recession—i.e., regardless of whether or not the countries were affected by the recession. On the other hand, they found that cultural background more consistently affects the capacity for innovation and competitiveness: pro-innovation societies always improve their innovation and competitiveness capacity, while anti-innovation societies reduce it, irrespective of macro-conditions.

Petrakis et al. (2014) used the performance of 24 European countries in terms of innovation and competitiveness during the Great Recession of 2008–2013 to clarify how much the change in innovation performance and competitiveness during 2008–2013 results from macroeconomic factors, versus more permanent factors such as cultural background. Their analysis identified two clusters of European countries: an "anti-innovation cluster" that includes countries with lower innovation activity and competitiveness and a "pro-innovation cluster" that includes countries with higher innovation and competitiveness. The pro-innovation culture cluster is characterized by an advanced cultural framework (sense of security and thus a greater tendency to take risks, higher general trust, creativity, and organizational structures that boost good performance), while the anti-innovation cluster possesses a cultural framework that hinders its innovative activity (reluctance to invest and use new technologies due to prevailing uncertainty, low interpersonal trust and hence reduced transactions, low creativity, and organizational structures that cannot innovate and hence increase organizational competitiveness). Countries in the same cluster seem to have similar cultural characteristics, followed by similarities in innovation index and competitiveness performance.

However, they conclude that the divergence of innovation and competitiveness of European countries depends not on macroeconomic environment but on culture. Culture plays a key role in improving the innovation performance of pro-innovation countries from 2008 to 2013, while macroeconomic environment does not—in fact a significant deterioration in macroeconomic environment is noticed during these years for the countries concerned. Simultaneously, the anti-innovation culture of the other cluster did not improve the performance of its constituent countries. Thus, by identifying the specific cultural characteristics of each society, we can also understand the differences in performance between countries. Real innovation is the result of open, rigorous activity and cooperation that cannot be continuously achieved. Owing to the lasting nature of culture, its effects on innovation and competitiveness levels can be observed only over the long term.

Cultural background can prevent or significantly promote economic growth, enabling macroeconomic conditions to play their role. Even if cultural background is challenged by severe macroeconomic conditions (as in Ireland), a society whose culture encourages innovation can still perform better in the future. In contrast, a society whose culture hinders innovation cannot achieve economic growth even if policy makers improve current macroeconomic conditions.

Thus, on the one hand, favorable conditions for the development of a climate of economic and political stability should prevail, by attracting innovative enterprises. For this purpose, handling uncertainty must be the main aim of policy makers. Investing in human capital is essential to disseminate knowledge to all members of society by creating specialized personnel able to deal with new technologies. The establishment of a proper institutional framework, including such aspects as property rights, rule of law, and transparency, can increase general trust. However, these policies may take 15–20 years to yield results.

On the other hand, managers should promote procedures for the collection, evaluation, and funding of innovative ideas that will create the conditions for an innovation culture. Simultaneously, teamwork is crucial, and that is why mutual trust should be built among team members. This is the concept of an innovation culture.

An efficient innovation culture should be able to identify key factors that drive innovation. People involved in innovation activities in an enterprise should not just focus on the results of specific scenarios but on how these results can be achieved. Appropriate leaders must exist who will have clarified to all participants the relevant targets and how to seek desired outputs. Focusing on results releases important energy that permits creativity.

Additionally, everyone has to be inspired since inspiration is the leadership competency that most profoundly influences productivity and engagement. When people are inspired by events or by a leader, they become incentivized to invest more effort.

Another important element is the challenge to the status quo, meaning there should be no prevailing sense of fear regarding the expression of views to senior executives of an enterprise or organization and, when needed, groups planning innovative activities should be able to disagree without being disagreeable. This fosters respect among their colleagues and generates creative discussions.

Other points that those involved in innovation should bear in mind are that: (a) high efficiency can kill innovation because it can stifle the creative culture necessary to foster innovation; (b) significant failures must be expected; not all ideas can prove successful, but the successful ones must offset the cost of the unsuccessful ones; (c) there should be an appropriate reward system for innovation activities; while formal rewards are good for the short term, they do not keep people truly engaged; (d) time is extremely important, innovations take time to develop, and proper time management enables more time to play with; additionally, providing employees "free" time to experiment with new technologies and products is very important to innovation; (e) there should be a balance between pressure and support; a new idea could fail because of a lack of adequate support from the enterprise; conversely, given excessive support, there is no pressure to move ahead; and (f) the focus should be on future challenges rather than past successes.

Acknowledgments This chapter is based on a research paper presented at the 3rd Global Innovation and Knowledge Academy (GIKA) Conference, 7–9 July 2014, Valencia, Spain, which was developed with the cooperation of P. C. Kostis and D. G. Valsamis. The author would like to thank Kyriaki I. Kafka, Pantelis C. Kostis, and Dionysis G. Valsamis (all researchers from the University of Athens) for their valuable contribution to the development of this chapter.

References

Almond, G. A., & Verba, S. (1965). *The civic culture: Political attitudes and democracy in five nations*. Boston: Little, Brown and Company.

Arrow, K. (1972). Gifts and exchanges. *Philosophy and Public Affairs, 1*, 343–362.

Baldwin, W. L., & Scott, J. T. (1987). Market structure and technological change. In *Fundamentals of pure and applied economics* (Vol. 17). Chur: Harwood Academic Publishers.

Bandura, A. (1963). *Social learning and personality development*. New York: Holt, Rinehart, and Winston.

Bandura, A. (1971). *Social learning theory*. New York: General Learning Press.

Barnett, H. G. (1953). *Innovation: the basis of cultural change*. New York: McGraw Hill.

Blackwell, S. S. (2006). The influence of perceptions of organizational structure and culture on leadership role requirements: the moderating impact of locus of control and self-monitoring. *Journal of Leadership & Organizational Studies, 12*(4), 27–49.

Burns, T., & Stalker, G. M. (1961). *The management of innovation*. London: Tavistock Publications.

Campbell, D., Stonehouse, G., & Houston, B. (2002). *Business strategy: An introduction*. New York: Routledge.

Cohen, W. (1995). Empirical studies of innovative activity. In P. Stoneman (Ed.), *Handbook of the economics of innovation and technological change*. Oxford: Blackwell.

Cohen, W. M., & Levin, R. C. (1989). Empirical studies of innovation and market structure. In R. Schmalensee & R. D. Willig (Eds.), *Handbook of industrial organization* (Vol. II). Amsterdam: North-Holland.

Davenport, T., DeLong, D., & Beers, M. (1998). Successful knowledge management projects. *Sloan Management Review, 39*, 43–57.

DeLong, D. W., & Fahey, L. (2000). Diagnosing cultural barriers to knowledge management. *Academy of Management Executive, 14*, 113–127.

Druker, P. (1968). *Frontiers of management*. New York: Truman Talley Books.

Foster, G. M. (1962). *Traditional cultures and the impact of technological change*. New York: Harper & Row.

Fukuyama, F. (1995). *Trust*. New York: Free Press.

Fukuyama, F. (1999, October 1) *Social Capital and Civil Society*. Paper presented at IMF Conference on Second Generation Reforms, The Institute of Public Policy, George Mason University.

Gellner, E. (1994). *Conditions of liberty: Civil society and its rivals*. London: Hamish Hamilton.

Halpern, D. (2005). *Social capital*. Cambridge: Polity.

Hana, U. (2013). Competitive advantage achievement through innovation and knowledge. *Journal of Competitiveness, 5*(1), 82–96.

Handy, C. (1993). *Understanding organizations*. London: Penguin.

Helfat, C. (1994). Evolutionary trajectories in petroleum firm R&D. *Management Science, 40*, 1720–1747.

Hofstede, G. (1980). *Culture's consequences: International differences in work-related values*. Beverly Hills, CA: Sage Publications.

Hofstede, G. (1984). *Culture's consequences: International differences in work-related values*. Beverly Hills, CA: Sage Publications.

Hofstede, G. (1991). *Cultures and organizations: Software of the mind*. London: McGraw Hill.

Holden, N. (2010). *Culture should be seen as a knowledge management resource. A vision on cultural diversity*. Rochester, NY: ADP publication.

House, R., Hanges, P., Javidan, M., Dorfman, P., & Gupta, V. (2004). *Culture, leadership and organisations – The GLOBE study of 62 societies*. Thousand Oaks, CA: Sage Publications.

Kamien, M. I., & Schwartz, N. L. (1982). *Market structure and innovation*. Cambridge: Cambridge University Press.

Knack, S., & Keefer, P. (1997). Does social capital have a economic payoff? A cross country comparison. *Quarterly Economic Journal, 112*, 1251–1288.

Nelson, R. R., & Winter, S. G. (1982). *An evolutionary theory of economic change*. Cambridge, MA: Harvard University Press.

Petrakis, P. E. (2014). *Culture, growth and economic policy*. New York: Springer.

Petrakis, P. E., & Kostis, P. C. (2012). The role of knowledge and trust in SMEs. *Journal of the Knowledge Economy, 15*(6), 105–124.

Petrakis, P. E., Kostis, P. C., & Valsamis, D. G. (2014). Innovation and competitiveness: Culture as a long-term strategic instrument. *Culture on Innovation and Competitiveness during the European Great Recession, 3rd Global Innovation and Knowledge Academy (GIKA) Conference*, Valencia, Spain.

Pohlmann, M. (2005). The evolution of innovation: cultural backgrounds and the use of innovation models. *Technology Analysis & Strategic Management, 17*(1), 9–19.

Putman, R. (1993). *Making democracy work: Civil traditions in modern Italy*. Princeton, NJ: Princeton University Press.

Rothwell, R., & Wissema, H. (1986). Technology, culture, and public policy. *Technovation, 4*(2), 91–115.

Shane, S. A. (1992). Why do some societies invent more than others? *Journal of Business Venturing, 7*, 29–46.

Shapero, A., & Sokol, L. (1982). Social dimensions of entrepreneurship. In C. A. Kent, D. L. Sexton, & K. H. Vesper (Eds.), *Encyclopedia of entrepreneurship*. Englewood Cliffs, NJ: Prentice Hall.

Snijders, T. A. B. (1999). Prologue to the measurement of social capital. *La Revue Tocqueville, 20*, 27–44.

Tocqueville, A. (1990). *Democracy in America I and II*. New York: Vintage.

Veblen, T. (1898). Why is economics not an evolutionary science? *Quarterly Journal of Economics, 12*(4), 373–397.

Chapter 10
Innovating Cultural Synergy in French–Russian Collaboration: Building a Puzzle from Cultural Diversity

Natalia I. Guseva

Abstract This chapter concerns an applied aspect of culture. Culture is regarded as the source for cultural diversity, as a reservoir for managerial knowledge, and as an opportunity to gain additional competitive advantages for global companies. Cultural distance between members of multicultural teams that is caused by the influence of the national culture is seen with a positive intent as an opportunity to achieve cultural synergy. The proposed approach for innovating cultural synergy is based on a three-stage model. The first stage is to define the most relevant features of the cultural diversity on an example of the French–Russian collaboration. During the second stage, the opportunities of cultural diversity are employed to create new knowledge reservoirs in the management process. The final third stage is to develop new creative managerial decisions and initiatives in order to achieve the synergy effect and to increase multicultural teams' management effectiveness.

Introduction

Globalization has become an objective reality today, and hence a new economic landscape dictates new rules and corresponding values. To stay competitive globally, more and more companies are increasing their activities and investments in foreign countries (Govindarajan and Gupta 2001; Solomon and Schnell 2009; Moran et al. 2011). The world is getting increasingly complex and diversified but, at the same time, more interconnected and unified (Martinelli 2003). For the last 15–20 years, the level of cultural diversity has been increasing tremendously. Nowadays, "cultural leverage" and management of cross-cultural differences are growing in importance as key issues of company management and as sources of competitive advantage.

N.I. Guseva, Ph.D. (✉)
National Research University Higher School of Economics, Moscow, Russia
e-mail: profguseva@gmail.com

I.N. Dubina, E.G. Carayannis (eds.), *Creativity, Innovation, and Entrepreneurship Across Cultures*, Innovation, Technology, and Knowledge Management, DOI 10.1007/978-1-4939-3261-0_10

In same period, the trend toward teamwork (Cohen and Mohrman 1995; Moore 1999; Salas et al. 2002, etc.) grew to one of the major organizational tendencies. Companies used teams as the most efficient form of human resource management. "Teams are in fashion," as observed by Thomas et al. (2000) to qualify the increasing prevalence of small groups and favorable conditions for employees' initiative as an extremely important prerequisite for the formation of competitive advantages of the company (Harris and Harris 1996; Bubshait and Farooq 1999; Hoegl et al. 2004).

Globalizing businesses and growing number of multinational companies require employees belonging to different cultures to work together: "Even in purely domestic operations, firms are being forced to form cross-functional, inter-departmental, cross-divisional, and inter-organizational alliances in order to make maximum use of scarce resources and thus increase their competitive advantage" (Shaw and Barrett-Power 1998). One of the major problems is that through globalization, companies are not given the choice of how they want to constitute their staff anymore (Maugain 2003).

The multicultural staff is becoming a common practice nowadays, which in turn has to be considered as a source for a sustainable competitive advantage or the reason for the emergence of multiple cross-cultural conflicts. One of the key contemporary research questions is the wide spectrum of the possible outcomes of multicultural teamwork. Lianne Roembke (2000) forwards a rhetorical question "Multicultural teams: a curse or blessing?". Even nowadays managing multicultural teams can be viewed as an exciting journey or "tightrope walk," depending on positive or negative company experiences (Maugain 2003). On the one hand, multicultural teams can turn into "extremely irksome stumbling blocks for a company or a project when not handled properly" (Dowling et al. 2008; Congden 2009). On the other hand, companies who recognize the potential of cultural diversity and find the right cultural compound of team members can harness corporate competitive advantages. Conducted research certifies the positive correlation between cultural diversity and innovativeness (Larsson and Finkelstein 1999). In this context, globalization offers much more tangible benefits for multinational companies.

Cultural Diversity and Creativity

Culture is an extremely complex intangible concept including different mind-sets guiding people in their problem-solving approaches and company behavior (Hofstede 1980; Hall 1992; Adler 1991; Trompenaars 1994; Schneider and Barsoux 1997; Lewis 2000, etc.). Leaning on various cultural models, "iceberg," "onion," "ocean," etc. (Hall 1976; Schein 1988; Trompenaars 1994; Chaney and Martin 1995; Schneider and Barsoux 1997), it has to be noted that "the way we really get things done" is different, and it inevitably leads to converse opinions, expectations, preferences, and interests of the people (Fig. 10.1).

Several spheres or types of culture influence every human being (Fig. 10.2). For instance, Barsoux and Schneider remark that "…these spheres constantly interfere with one another, so that people do not have one single culture, but are carried by a

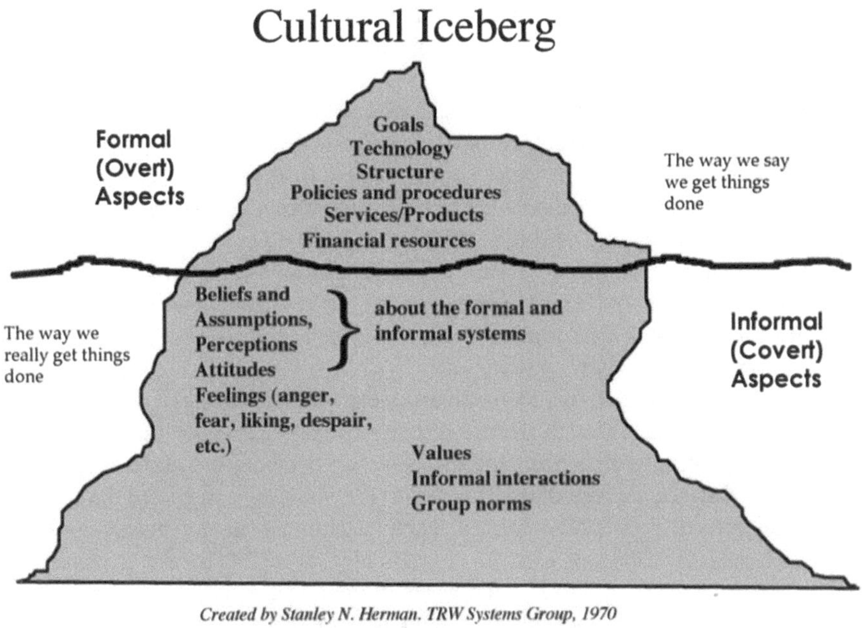

Fig. 10.1 Hall cultural iceberg model

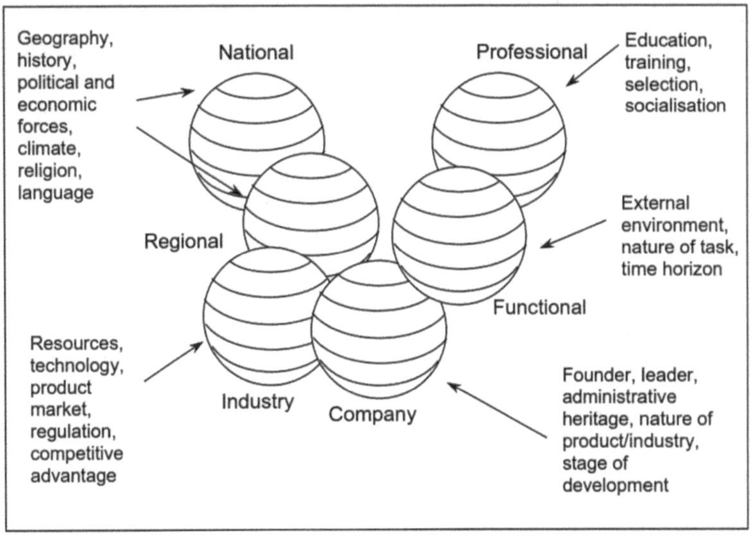

Fig. 10.2 Interacting cultural spheres of influence (*Source*: Schneider and Barsoux 1997)

nexus of national, regional, industry, company, functional and professional cultures" (Schneider and Barsoux 1997; 2003). Therefore, it could be viewed as a cultural diversity.

It can be stated that culture and diversity are two interrelated and complementary notions. Differences in national, corporate, professional cultures inevitably lead to diversity within interactions between team members. Besides, differences in cultural background among team members lead to differences in value systems, behavioral norms, views, experience, and skills, which cause diversity.

Diversity can be generally described as "the condition of being different or having differences" (Merriam-Webster 1986). Given that the research focuses on workforce diversity, the following definitions present the notion in an organizational context ranging from broad to precise descriptions.

According to Byeong Yong (2006), diversity can be considered as a wide spectrum of variation that distinguishes the human resources of an organization. Jackson argued that the term diversity recognizes the fact that any company's workforce includes people from different backgrounds (Jackson 2005): "Diversity is created through the differences among people who form a team, department, or organization; consequently it does virtually not exist at the individual level." "Diversity," confirmed Davie, "refers to the heterogeneity of a group or organization based on the inclusion of individuals of different backgrounds and experiences" (Davie 2005).

Besides rephrasing diversity more accurately as a mixture of people with different social and demographic group identities within the same social system, Nkomo and Stewart also enlarged the meaning of diversity as to how differences in human resources influence social relations at the workplace (Nkomo and Stewart 2006). In contrast to most diversity definitions, Thomas not only included the differences but also the similarities between and among the workforce into the concept of diversity (Thomas 1999).

Diversity definitions also take into consideration that "there are differences even within a particular group" (Ollapally and Bhatnagar 2009). Diversity, furthermore, is the variety arising out of differences and similarities in intellectual capabilities (Bassett-Jones 2005); qualities, perspectives, and attitudes (Robinson and Dechant 1997); and "beliefs, understandings, values and ways of viewing the world" (Shen et al. 2009).

Thomas suggested a definition that represents a relatively holistic approach to capture the concept of diversity: "the amount of cultural heterogeneity represented in a team." He stated that, "…apart from social, demographic and intrapersonal variables of diversity, the interpretation extends to work-related variables like corporate background, function, tenure with the organization, exempt and non-exempt status, and management and non-management…"(Thomas 1999).

Cultural diversity affects all major management processes such as communication and team building, empowerment and leadership, problem solving and decision making, resolving conflicts and negotiating, as well as motivating members of a multicultural team. While cooperating, members of multicultural teams face the fact that cultural diversity can induce controversial answers and reactions to the same set of management practices and processes (Harrison et al. 2000). Therefore, cultural diversity generally leads to numerous conflicts, due to differences in employees' value systems and decisions based on them. At same time, there is a positive side of cultural diversity. It increases new idea generation and provides development of unconventional and innovative decisions, opportunities for cultural synergies, and the creativity in business in general.

Amabile pointed out three major components for the creativity enhancement in business (Amabile 1998; Amabile and Kramer 2010; 2012). These are expertise, creative thinking, and motivation. The researcher suggests to consider a combination of technical, procedural, and intellectual capabilities as expert knowledge. Creative thinking skills demonstrate "how people approach problems and solutions—their capacity to put existing ideas together in new combinations. The skill itself depends quite a bit on personality as well as on how a person thinks and works." Motivation is viewed from a dominant factor perspective, either intrinsic or extrinsic. Intrinsic motivation, which is in our particular interest, is based on the internal value system of an employee and can appear as creativity, satisfaction from accomplished tasks, self-revealing, self-realization, etc.

In order to maximize the creativity effect in business, a sufficiently high degree of cultural diversity and efficient collaboration between team members is required. The magnitude of these differences depends on two groups of factors. Firstly, it depends on the level of cultural distance between representatives of different cultures and, secondly, on the number of cultures represented in a multicultural team (Thomas et al. 2000). In summary, the higher the magnitude of cultural diversity, the more relevant the issues of cultural synergy achievement on cultural clashes in a global environment are.

Concept of Cross-Cultural Synergy

Cross-cultural synergy is becoming one of the most innovative and effective management concepts for companies operating in the global environment. The power to combine the perspectives, resources, and skills of a group of people and organizations is called synergy (Fried and Randall 1994; Lasker et al. 1997; Richardson and Allegrante 2000). Lasker and colleagues identify synergy as "the proximal outcome of partnerships that, in turn, influences the effectiveness of a partnership" (Lasker et al. 2001).

Synergy, a term that first appeared in Greek, means *working together* (*sun*, *together*; *ergon*, *working*). Synergy implies a belief that we can learn from each other. Thus, for instance, a leadership model based on cultural synergy principles can be described as a combination of Chinese wisdom, German formalism, French charm, Dutch originality, and Russian glamour. Consequently, cultural differences can lead to mutual growth and achievement, with a higher synergy effect, compared to a simple sum of contributions of each party.

Cultural synergy is a dynamic approach to managing cultural diversity in a variety of contexts. Cultural synergy builds on common ground, transcending mere awareness of difference, to form models of effective communication, leadership, decision making, versatile strategic alliances and partnerships, etc.

Synergy could be achieved on micro- and macro-levels. It starts between individuals, then extends to their teams and organizations, and finally involves the whole societies. Cultural synergy is a creative process and presents a very powerful concept based on adapting and learning from each other. Thus, synergy could be reached on knowledge co-creation. It doesn't signify any compromise as nothing is given up or lost in the process of *true synergy*.

Multinational organizations have a special role not only in building cross-cultural bridges but in innovating synergies through their practical knowledge of putting together human and natural resources with the know-how of managing both in the most effective ways.

Moran et al. (2011) emphasize that synergy involves joint team member action in which the total effect is greater than the sum of effects when acting independently; it also creates an integrated solution and, last but not least, develops the potential of members by facilitating the release of team energies. The researchers stated that practice of synergy capitalizes on difference through cooperation and collaboration: "…Global leaders who are dedicated to accessing the benefits of globalization know that multilateralism with a commitment to cultural synergy as a tool helps to empower all people of multicultural, diverse backgrounds…."

Lasker and colleagues made an attempt to operationalize partnership synergy (Lasker et al. 2001). They described it as the extent to which the contributions of different partners improve the ability of the partnership to:

- Think about its work in creative, holistic, and practical ways.
- Develop realistic goals that are widely understood and supported.
- Plan and carry out comprehensive interventions that connect multiple programs, services, and sectors.
- Understand and document the impact of its actions.
- Incorporate the perspectives and priorities of community stakeholders, including the target population.
- Communicate how its actions will address community problems.
- Obtain community support.

So, the cross-cultural synergy that a partnership achieves is reflected in the way stakeholders think about the partnership's goals, plans, and evaluation; the types of actions the partnership carries out; and the relationship the partnership develops with the broader community (Butterfield et al. 1996; Roussos and Fawcett 2000).

The innovating synergy is possible under the condition of employees' creativity, when members of the team are seeking new ways to solve problems, make decisions, and demonstrate their capabilities of creating a "new picture"—a puzzle with interrelated elements, which, once combined in a new way, generate synergy. We called cultural synergy as a puzzle, built of separate elements of cultural diversity. Each of them is unique and relevant, has a certain impact, and is connected to other elements.

To sum up, the factors that enhance synergy are creativity, comprehensive thinking that enables to see the problem as a whole, practical thinking, and transformative thinking. Moreover, complementarity of partners is very important (Lasker et al. 2001). Synergy could be considered as management innovation based on the cross-cultural bridges and putting together different perceptions, knowledge, practical skills, and competencies and experiences. Cross-cultural synergy is a product of culture interaction within the group and can be assessed in concrete, practical ways.

French–Russian Cultural Diversity

In order to measure cultural diversity in the framework of French–Russian collaboration and see the potential for innovating synergies, we have used an integrated multi-method analysis of cultural differences, which influences management theory and practice. Methodological research done by Kluckhohn and Strodtbeck (1961), Hofstede (1980, 1991, 1995, 2001, 2010), Triandis (1972, 1994, 1996), Hall (Hall and Hall 1990; Hall 1992), House (House et al. 2004), Trompenaars and Hampden-Turner (Trompenaars 1994; Hampden-Turner and Trompenaars 2004), Schneider and Barsoux (1997, 2003), and d'Iribarne and colleagues (1998) have formed the theoretical model outlining cultural dimensions (Table 10.1).

The results of the French–Russian analysis with particular focus on cultural diversity, based on different cross-cultural models, allowed us to identify *three groups of French–Russian cultural differences* (Guseva 2004a, b; 2011). The main interest and research focus were given to the first group that consisted of the most significant cross-cultural differences, including four fundamental dimensions of culture:

- Universalism–particularism
- Individualism–collectivism
- Endogenous–exogenous motivation
- Attitudes toward time

In four identified cultural dichotomies between Russian and French, there are two extremes that can be found, which allow the distinction of all options for innovating cultural synergy, moving from extreme positions and learning from each other. Each pair of cultural settings is one another's reflection. There is no reason to say which approach of solving problems or management practice or way of doing business is better (Adamopoulos 1999; Harrison and Huntington 2000; Hampden-Turner and Trompenaars 2004). Individualism is a reflection of collectivism, particularism is a reflection of universalism, and vice versa. However, between each pair of parameters that characterize national culture, there is a sort of a "vicious circle." It is therefore necessary to find a compromise, a kind of a middle ground, then create a harmony, and develop synergy from cultural diversity.

Cross-cultural differences mentioned above play an important role for all major management processes such as communication and multicultural team building, empowerment and leadership, problem solving and decision making, resolving conflicts and negotiating, as well as motivating members of a multicultural team. We validate these ideas using survey data on 51 foreign-owned companies operating in Russia and then present case studies designed to justify the results for the Russian

Table 10.1 Major concepts of cross-cultural research

Concepts	Dimensions of national culture								
1. The paradigm of "value orientations" by Hofstede	Individualism–collectivism	Power distance	Masculinity–femininity	Uncertainty avoidance	Indulgence–restraint	Monumentalism	Long-term/short-term orientation		
2. Conceptual scheme of "individualism–collectivism" by Triandis	Individualism–collectivism	Complexity–simplicity		Reglamentation–uncertainty					
3. Method (modern dilemmas) by Trompenaars and Hampden-Turner	Individualism–collectivism	Universalism–particularism	Specific–diffuse	Achievement–ascription	*Neutral–emotional*	Sequential and simultaneous attitudes toward time	Exogenous–endogenous motivation		
4. Conception (cultural grammar) by Hall			Speed of information flow		Low context–high context	Monochrome–polychrome	Small and large personal space		
5. GLOBE project by House	Collectivism-I	Collectivism-II	Power distance	Uncertainty avoidance	Gender egalitarianism	Assertiveness	Performance orientation	Future orientation	Humane orientation

context. The results of the study, based on a combination of quantitative and qualitative methods, allowed us to determine the "Russian" and the "French" efficient way of management in the context of cross-cultural differences in core management processes and propose the mechanism for creating synergy.

French–Russian Partnership Synergy

In order to innovate cultural synergy within the framework of French–Russian collaboration, the paper provides a model based on a three-stage approach. The first stage serves to define the most relevant features of the cultural diversity on an example of the French–Russian collaboration. During the second stage, the opportunities of cultural diversity to create new "knowledge reservoirs" in the management process are highlighted. The final third stage helps to develop new creative managerial decisions and initiatives in order to achieve the synergy effect and to increase multicultural teams' management effectiveness.

Partnership synergy can be based on four major French–Russian cultural dichotomies: universalism–particularism, individualism–collectivism, endogenous–exogenous motivation, and attitudes toward time. In this paper, we have focused on the first one: universalism–particularism.

In relation to the French, Russians are a more particularistic culture. Consequently, despite their intermediate position on the scale of universalism–particularism, the French will be more universalistic representatives, which in turn can be a source of cross-cultural conflicts in a multicultural team or a source of synergies.

For instance, although the French concentrate on relationships, they are nevertheless a nation of individualists. Moreover, in spite of their unwillingness to stick to the subject and conversations frequently going "as the situation may demand," the French are easily involved in arguments and disagree in a sharp form, mincing no words, at the negotiation desk. And, despite the word "egalitarian" coming from "égalité"—"equality"—France now remains one of the most hierarchical nations in Europe. In other words, French leaders are prone to focus on relationships and are nonetheless individualists, with a fine perception of the situation, and context oriented. It is a very unusual combination of cultural features.

According to the results of our empirical studies, major French–Russian cultural differences in management processes have been distinguished (Table 10.2).

In order to achieve the synergetic effect while doing business, interacting in a team, and negotiating, it is required to take the opportunities from French–Russian cultural differences and make advantage of the partner's cultural feature awareness (Table 10.3).

Russia represents a "low synergetic" society based on Benedict and Maslow approach with lack of synergetic relations in the culture. It represents a highly complex and multicultural society with differing and conflicting beliefs, values, religions, and cultures of communication between each other. It also means uncooperative, individualistically oriented, and aggressive behavior with "win–lose" attitudes between members.

Table 10.2 French–Russian cultural diversity

Russians	French
1. Assign greater importance to the relationship with a colleague/partner than to the existing rules and regulations	1. Assign greater importance to the existing rules and regulations than to the relationship with a colleague/partner
2. A contract is the beginning of business relations with the partner	2. A contract is the beginning of a concrete project
3. Contracts are flexible and constantly changing. Greater attention is paid to continuous adjustments on the agreement	3. Contracts keep their initial form. Greater attention is paid to completeness and universality
4. A reliable business partner or employee is the one who is ready to make concessions if the partnership terms change	4. A reliable business partner or employee meets obligations (contract terms) and has respect for his/her own promises
5. What has been said can be trusted	5. Only what is written can be trusted
6. Nonverbal communication matters	6. Verbal communication is enough

Table 10.3 Opportunities for French–Russian partnership synergy

Synergy with Russians	Synergy with the French
1. Special attention has to be paid to form personal relationships with colleagues/partners. Do not consider conversations on personal subjects to be useless	1. Do not consider the intention of the French to focus on business and problem solving and lack of interest to your personal problems rude or offensive
2. Prepare a framework agreement for cooperation, which will be the basis for the partnership	2. Be prepared to clearly identify and register all conditions of cooperation, including the objectives, activities, participants, and deadlines
3. You must be prepared for uncertainty, various comments, or suggestions that may be of no value	3. You must be prepared for rational and professional discussions and argumentation of proposals and focus on achievement
4. Carefully preserve your legal safeguards	4. You should carefully examine all the legal issues with a lawyer if you have any doubts
4. Be more flexible and ready to make concessions on various issues	4. Determine the initial conditions appropriate for cooperation
6. It is necessary to conduct important negotiations in the informal atmosphere on the basis of pre-built warm/friendly relationships with colleagues and partners	6. Considering the results of the agreement achieved with a potential partner, a letter of intent should be signed or an e-mail sent confirming the main points of the agreement

France is a "middle-level" synergetic society and individualistically oriented, with a low capacity for cooperation on the organizational level, based on conflicting values and beliefs with moderate "win–win" attitudes toward colleagues and partners. The French are a peculiar sort of people with a unique culture of doing business. It is unique due to the combination of both the ancient German influence of the north of Europe and the Latin infusion from the south.

During the process of multicultural team management, the main focus should be set on mutual understanding, openness, and readiness of team members to changes (see

Table 10.4 Opportunities by developing new initiatives and activities in French–Russian multicultural team

Managerial recommendations for Russians	Managerial recommendations for the French
1. Informal networks of contacts and connections should be formed, creating an atmosphere of personal understanding, sensitivity, and responsiveness	1. Make effort to be consistent and aim toward uniformity of procedures
2. Attempt to informally change habitual patterns of behavior	2. Establish formal ways to change the existing business methods and practices
3. Adjust relationships with you, in order for you to change the system	3. Adjust the system for the system to modify you
4. Use implicit controls	4. Explicitly declare change
5. Seek justice, examining each individual case substantially, taking into account the extent of its merits	5. Seek justice, treating all similar cases equally

Table 10.4). Results of empirical studies showed that 97.3 % of the French and 88.9 % of the Russians have strong desire to continue the French–Russian collaboration; hence, synergy of cross-cultural differences is possible (Guseva 2011).

So, for Russians fundamental shifts should be done on joint decision making, openness to change, innovation, group consensus, and creative problem solving. Taking into account French goals—directed approach and high team performance—there is strong necessity to clarify roles, relationships, and responsibilities in the French–Russian team.

Conclusion

Globalizing business leads to an increasing number of multicultural organizations operating in Russia. Within the Russian context, global companies attempt to apply management approaches and management practices being successful in their own cultural environments. However, many conflicts have arisen due to fundamental differences on cultural values, sharing perceptions, experiences, and management practices, which undermines performance and productivity. In this regard, the most topical issue is the use of cultural diversity for innovating cultural synergy for additional competitive advantage.

Cultural synergy can be defined as the extent to which the multinational companies' perspectives, resources, and skills of participating individuals and organizations contribute to and strengthen the teamwork in company activities. The process of encouraging cross-cultural synergy between the French and Russians is based on managing the impact of cultural diversity and cultural conflict resolution by sharing culture values, patterns, knowledge, and management experiences. It unites team members with complementary talents, dissimilar demographic characteristics (age, race, gender, etc.), and different cultures.

Synergetic approach to multicultural team management harnesses future MNC's opportunities by developing new initiatives, activities, and innovative ideas; arises interest and creates preconditions for creativity; urges an atmosphere of constructive criticism; cultivates listening skills; and fosters core values such as trust, confidence, and commitment within French–Russian teams.

References

Adamopoulos, J. (1999). The emergence of cultural patterns of interpersonal behavior. In *Social psychology and cultural context* (pp. 63–76). Thousand Oaks, CA: Sage.

Adler, N. (1991). *International dimensions of organizational behavior* (2nd ed.). Boston: Kent Publishing Company.

Amabile, T. M. (1998, September–October). How to kill creativity. *Harvard Business Review*, pp. 76–87.

Amabile, T. M., & Kramer, S. J. (2010). What really motivates workers (#1 in breakthrough ideas for 2010). *Harvard Business Review, 88*(1), 44–45.

Amabile, T., & Kramer, S. (2012, January). How leaders kill meaning at work. *McKinsey Quarterly*.

Bassett-Jones, N. (2005). The paradox of diversity management, creativity and innovation. *Creativity and Innovation Management, 14*(2), 169–175.

Bubshait, A., & Farooq, G. (1999). Team building and project success. *Cost Engineering, 41*(7), 34–38.

Butterfield, F., Goodman, R., & Wandersman, A. (1996). Community coalitions for prevention and health promotion: factors predicting satisfaction, participation, and planning. *Health Education Quarterly, 23*, 65–79.

Byeong Yong, K. (2006). Managing workforce diversity: developing a learning organization. *Journal of Human Resources in Hospitality & Tourism, 5*(2), 69–90.

Chaney, L., & Martin, J. (1995). *Intercultural business communication*. Englewood Cliffs, NJ: Prentice Hall.

Cohen, S., & Mohrman, S. (1995). *Designing team-based organizations*. San Francisco: Jossey-Bass.

Congden, S. (2009). Cross-cultural communication and multicultural team performance: a German and American comparison. *Journal of Comparative International Management, 12*(2), 73–89.

d'Iribarne, P., Henry, A., Segal, J.-P., Chevrier, S., & Globokar, T. (1998). *Cultures et mondialisation*. Paris: Seuil.

Davie, S. L. (2005). Diversity. In P. H. Werhane & R. E. Freeman (Eds.), *The Blackwell encyclopedia of management* (Business ethics 2nd ed., Vol. 2, pp. 141–147). Malden, MA: Blackwell Publishing.

Dowling, P., Festing, M., & Engle, A. (2008). *International human resource management: Managing people in a multinational context* (5th ed.). London: South-Western.

Fried, B. J., & Randall, T. G. (1994). Managing groups and teams. In S. M. Shortell & A. D. Kaluzny (Eds.), *Health care management organization, design, and behavior* (3rd ed.). Albany, NY: Delmar.

Govindarajan, V., & Gupta, A. (2001). Building an effective global business team. *Sloan Management Review, 42*(4), 63–71.

Guseva, N. (2004a). *Efficient management of multicultural teams: Methodological and applied aspects*. Irkutsk: BGUEP.

Guseva, N. (2004b). Theoretical and methodological approaches of Franco-Russian cross-cultural research. In *Proceedings of EMCSR 2004*. Vienna: Vienna University Press.

Guseva, N. (2011). Elements of strategy of French – Russian efficient cooperation: cross-cultural approach. In *Modern management: Problems, hypothesis, research* (pp. 46–57). Moscow: HSE & Harvard Business Review Pub.

Hall, E. (1976). Beyond Culture. New York: Anchor Books.

Hall, E. (1992). *Dance de la vie: Temps culturel, temps vecu*. Paris: Le Seuil.

Hall, E., & Hall, M. (1990). *Understanding cultural differences – Germans, French and Americans*. Yarmouth, ME: Intercultural Press.

Hampden-Turner, C., & Trompenaars, F. (2004). *Au-dela du choc des cultures: Depasser les oppositions pour mieux travailler ensemble*. Paris: Editions d' Organisation.

Harris, P., & Harris, K. (1996). Managing effectively through teams. *Team Performance Management: An International Journal, 2*(3), 23–36.

Harrison, L., & Huntington, S. (2000). *Culture matters: How values shape human progress*. New York: Basic Books.

Harrison, G., Mckinnon, J., Wu, A., & Chow, C. (2000). Cultural influences on adaptation to fluid workgroups and teams. *Journal of International Business Studies, 31*(3), 489–505.

Hoegl, M., Weinkauf, K., & Gemuenden, H. (2004). Interteam coordination, project commitment, and teamwork in multiteam R&D projects: a longitudinal study. *Organization Science, 15*, 38–55.

Hofstede, G. (1980). *Culture's consequences: International differences in work-related values*. Beverly Hills, CA: Sage Publications.

Hofstede, G. (1991). *Cultures and organizations: Software of the mind*. London: McGraw Hill.

Hofstede, G. (1995). Motivation, leadership and organization: do American theories apply abroad? In C. Bartlett & S. Ghoshal (Eds.), *Transnational management: Text, cases and readings in cross-border management*. Chicago: IRWIN.

Hofstede, G. (2001). *Culture's consequences: Comparing values, behaviors, institutions, and organizations across nations*. Thousand Oaks, CA: Sage Publications.

House, R., Hanges, P., Javidan, M., Dorfman, P., & Gupta, V. (2004). *Culture, leadership and organisations – The GLOBE study of 62 societies*. Thousand Oaks, CA: Sage Publications.

Jackson, S. E. (2005). Diversity. In L. H. Peters, C. R. Greer, & S. A. Youngblood (Eds.), *The Blackwell encyclopedia of management, Vol. V: Human resource management* (pp. 94–95). Malden, MA: Blackwell Publishing.

Kluckhohn, F., & Strodtbeck, F. (1961). *Variations in value orientations*. Evanston, IL: Row, Peterson.

Larsson, R., & Finkelstein, S. (1999). Integrating strategic, organizational, and human resource perspectives on mergers and acquisitions: a case survey of synergy realization. *Organization Science, 10*, 1–26.

Lasker, R.D. & The members of the committee on medicine and public health (1997). Medicine & public health: the power of collaboration. Chicago: Health Administration Press.

Lasker, R. D., Weiss, E. S., & Miller, R. (2001). Partnership synergy: a practical framework for studying and strengthening the collaborative advantage. *The Milbank Quarterly, 79*(2), 179–205.

Lewis, R. (2000). *When cultures collide – Managing successfully across cultures*. London: Brealy Publishing.

Martinelli, A. (2003). Markets, governments, communities and global governance. *International Sociology, 18*, 291–323.

Maugain, O. (2003) *Managing multicultural R&D teams - An in-depth case study of a research project at CERN*. Dissertation Nr. 2820. St. Gallen University.

Merriam-Webster. (1986). *Webster's third new international dictionary of the English language unabridged* (Vol. I). Chicago: Encyclopaedia Britannica.

Moore, C. (1999). Intra-team communication. In R. Stewart (Ed.), *Gower handbook of teamworking* (pp. 210–223). Aldershot: Gower.

Moran, R., Harris, P., & Moran, S. (2011). *Managing cultural differences* (8th ed.). Oxford: Elsevier.

Nkomo, S. M., & Stewart, M. M. (2006). Diverse identities in organizations. In S. R. Clegg, C. Hardy, T. B. Lawrence, & W. R. Nord (Eds.), *The SAGE handbook of organization studies* (2nd ed., pp. 520–540). Thousand Oaks, CA: Sage Publications.

Ollapally, A., & Bhatnagar, J. (2009). The holistic approach to diversity management: HR implications. *Indian Journal of Industrial Relations, 44*(3), 454–472.

Richardson, W., & Allegrante, J. (2000). Shaping the future of health through global partnerships. In C. E. Koop, C. E. Pearson, & M. R. Schwartz (Eds.), *Critical issues in global health*. San Francisco: Jossey-Bass.

Robinson, G., & Dechant, K. (1997). Building a business case for diversity. *Academy of Management Executive, 11*(3), 21–31.

Roembke, L. (2000). *Building credible multicultural teams*. Pasadena, CA: William Carey Library.

Roussos, S., & Fawcett, S. (2000). A review of collaborative partnerships as a strategy for improving community health. *Annual Review of Public Health, 21*, 369–402.

Salas, E., Bowers, C., & Edens, E. (2002). *Improving teamwork in organizations: Applications of resource management training*. Mahwah, NJ: Lawrence Erlbaum.

Schein, E. (1988). Does Japanese management style have a message for American managers? In R. Katz (Ed.), *Managing professionals in innovative organizations – A collection of readings* (pp. 370–383). New York: Harper Collins Publishers.

Schneider, S., & Barsoux, J.-L. (1997). *Managing across cultures*. Harlow: Prentice Hall.

Schneider, S., & Barsoux, J.-L. (2003). *Management intercultural* (2nd ed.). Paris: Pearson Education Ltd.

Shaw, J., & Barrett-Power, E. (1998). The effects of diversity on small work group processes and performance. *Human Relations, 51*(10), 1307–1325.

Shen, J., Chanda, A., D'Netto, B., & Monga, M. (2009). Managing diversity through human resource management: an international perspective and conceptual framework. *International Journal of Human Resource Management, 20*(2), 235–251.

Solomon, C. M., & Schnell, M. C. (2009). *Managing across cultures: The 7 keys to doing business with a global mindset*. New York: McGraw-Hill.

Thomas, D. (1999). Cultural diversity and work group effectiveness. *Journal of Cross-Cultural Psychology, 30*(2), 242–261.

Thomas, D. C., Ravlin, E. C., & Barry, D. (2000). Creating effective multicultural teams. *University of Auckland Business Review, 2*(1), 11–24.

Triandis, H. C. (1972). The analysis of subjective culture. New York: Wiley-Interscience.

Triandis, H. C. (1994). Culture and social behavior. New York: McGraw-Hill.

Triandis, H. (1996). The psychological measurement of cultural syndromes. *American Psychologist, 51*, 407–415.

Trompenaars, F. (1994). *Riding the waves of culture: Understanding cultural diversity in business*. Chicago: Irwin Professional Publishing.

Chapter 11
A Cultural Context, Curiosity, and Creativity of Innovators: Marie Curie, Nikola Tesla, and Steve Jobs

Phillip S. Harvard

Abstract Comparing the lives of three very creative people—Marie Curie, Nikola Tesla, and Steve Jobs—this chapter provides an indicative comparative analysis, though not representative, on how their curiosity cultivated their creativity within different cultural contexts.

To Begin with…

Anthropologists usually see habits, traditions, ceremonies, and rituals as essential elements in defining a culture. Everyday dressing, eating, housing, working, playing, and learning become parts of the repeated habits, traditions, ceremonies, and rituals composing a given culture. Historically speaking, certain people in any culture always seem to have willingly learned how to work better than others allowing them the pleasure of dressing, eating, being housed, and playing, while others are obliged to keep on working in order to dress, eat, and be housed before being able to play at all. Leaving behind twentieth capitalistic open markets of free trade based on competition, I propose focusing on the way people have lived their culture and why only certain, and not others, have cultivated their curiosity to the level of creativity that can lead to innovations. Can curiosity flourish as a way of learning under contrasting conditions? Can it flourish according to different cultures? Does mother culture determine curiosity and creativity?

Comparing the lives of three very creative people furnishes information to permit an indicative comparative analysis, though not representative, on how their curiosity, within different cultural contexts, cultivated their creativity until it bore the ingenious fruits for which we honor, respect, and remember them: Marie Curie, Nikola Tesla, and Steve Jobs.

P.S. Harvard, M.B.A., D.E.A., D.E.S.S. (✉)
EIGSI General Engineering School, La Rochelle, France
e-mail: phillip.harvard@eigsi.fr

I.N. Dubina, E.G. Carayannis (eds.), *Creativity, Innovation, and Entrepreneurship Across Cultures*, Innovation, Technology, and Knowledge Management, DOI 10.1007/978-1-4939-3261-0_11

Marie Curie was a highly intelligent and very disciplined tenacious woman and a devoted wife and mother. Nikola Tesla was a mind beyond his time. Steve Jobs was an abandoned adopted child with no university degrees yet revolutionized today's ICT (Information and Communication Technology) industry. A precise purposeful peek through the keyholes of their lives can only unlock new insights to enlarge our understanding of the link between culture, curiosity, and creativity. How did they live? Who were they? What did they study? What were their working conditions? Why did they create?

Ladies First...

Maria Salomea Sklodowska was born in 1867 in Warsaw and was the youngest of five children. Her father lost his position teaching physics and math because of his patriotic tendencies under the Czarist control of Poland. The Polish language was to be stamped out to keep the Poles ignorant of their history and culture. Her father's subsequent series of unstable teaching jobs made family life materially difficult, and rooms were rented out. It was from one of these boarders her mother contracted typhus and died after an anguishing 5-year battle with tuberculosis. Having already lost a sister the same way, Maria turned to her studies and let curiosity become her antidote to escape everyday worry and hurt. The widower and his children were a close strong family unit, and on Saturday nights he would read to them the classics and taught them physics with an original experimental apparatus he was forbidden to use by his former Russian secondary school director.

Manya, as her father called her, excelled and was a brilliant student receiving her secondary degree at the age of 15 with highest honors and a medal. Afterward, she was sent by her father for a year to visit country cousins and recuperate from depression, or a burnout, due to overdoing and overworking a 1000 % to escape that which was unpleasant in her life. Thus, this future pattern of her later life had already taken root early on in her youth. Recognition, status, and identity became her motivation through scientific achievement due to tough, tedious, and tenacious hard work.

Bronia, an older sister, went to Paris to study medicine and married a fellow Polish medical student, Casimir Dluski. Manya first lodged in Paris with Bronia and Casimir while completing a university degree distinguishing herself with honors as number one in her graduating class of 1892 at the Sorbonne. In her days, only 23 (mostly foreigners) of the students at the Sorbonne were women. She received a scholarship to pursue a second university degree, not in physics this time, but in mathematics and was second not first, in her graduating class. Although Manya now lodged in the Latin Quarter, her sister did her best to make Manya eat after fainting from overstudying and by having her wear all the clothes she owned just to stay warm in winter. Her poor technical French and nonuniversity level in science and math were not stumbling blocks for Manya but steps to climb higher. Once again, the intellectual exercise of scientific achievement remains the motivation to be curious and earn recognition through degrees. Very patriotic and loyal to her country and her culture, she planned to return her family in Poland the following year but stayed in France to accept the invitation of a respected French scientist, Gabriel Lippmann, to work in is his lab where she was commissioned to do research about magnetism of metals.

Her new network of scientists included a fellow Pole from the University of Fribourg who introduced her to a homologue also doing research about magnetism, Pierre Curie, from the French School of Physics and Chemistry. Finally, she actually did return home to teach and contribute to Polish emancipation against Russian oppression, but Pierre Curie's letters seduced her to return and pursue a doctorate. Back in France once again, she pursued a doctoral thesis organizing her research notes about magnetism of metals into readable traceable documents. She encouraged Pierre to pursue his doctorate which gave him a higher teaching salary in 1895. Pierre convinced her their common passion for scientific research guaranteed their friendship that had blossomed into love, and they had a nonreligious civil marriage ceremony without rings that same year at the town hall of Sceaux, France. The navy blue dress she wore at the wedding proved to be the ideal work garment for the new Marie Curie as she continued her research, by invitation, under inadequate and rudimentary conditions in the lab of her former professor and mentor, Lippmann, who kept finding her sponsors hence financing. She focused all her time and energy on her doctoral research about the rays of Becquerel and discovered other elements more radioactive than uranium even four times (Polonium) more radioactive and up to nine times (radium) more. She discovered radioactivity was a property linked to atoms not a chemical property. She used an instrument of measurement for rays invented by her husband, Pierre, and his brother, Jacques. Pierre put aside his own research and began assisting his wife.

A German scientist visited her lab, and at first thought it was a joke to mock him, until he realized the cold cruel reality of her miserable working conditions. He could not believe his eyes, thinking at first it was just a miserable potato barn. In 1897, Pierre and Marie's first daughter was born, Irene. The next year Marie was awarded, for the first of three times (1898/1900/1902), the Gegner Prize meaning 4000 Francs each time toward her research. A second daughter, Eve, was born in 1903, the year Marie defended her doctoral thesis. Yet an even greater recognition was given that year as she shared the Nobel Prize of Physics with her husband and Henri Becquerel. In the following year, Pierre passed away, and she found much needed comfort in his family benefiting from the invaluable support of Pierre's father and brother. Marie replaced Pierre teaching his classes, and it was in 1907 that she moved her lab to Sceaux to be closer to where he was buried. She continued her research about radioactivity and isolated radium as a pure metal before writing her famous essay about radioactivity. The Pasteur Institute proposed setting up her own Institute of Radium on their premises. The rise of antifeminist press coverage against Marie Curie was xenophobic and myogenic and resulted in her being refused membership in the French Institute. As a young widow of 38 years, Marie's friendship with a fellow French scientist, Paul Langevin, was perverted by the press into a scandal, and she lost her nomination as the first woman to the famous French Academy by only two votes. The French government even requested Marie to leave France definitively and return to her home country of Poland.

She was the only woman invited to participate at the famous 1911 Solvay Scientific Congress in Belgium in the presence of Einstein, Poincare, Planck, and Rutherford. The same year she received a second Nobel Prize, this time in chemistry, for her discovery of the two elements: polonium and radium. She followed Einstein's advice and ignored the slanderous press and went to Stockholm for her Nobel Prize while the press remained silent. Einstein was later known to have said Marie Curie was the only uncorrupted famous person he ever met. Following her lifelong pattern, again she lost herself in her research and worked even more than ever before leading up to her work about atoms. During World War I, Marie, assisted by her older daughter, Irene, went to the front lines providing mobile surgery units known as *petites curies* to help the wounded. She actually built the prototype for these mini-ambulances using her friend, the Princess of Polignac's car, and the lab equipment of a fellow research colleague. Her radiology allowed identifying more precisely the location of bullets and shards of metal from explosives before operating to remove them out of the bodies of soldiers. She trained over 150 women radiology technicians in her Institute of Radium for the military hospitals. It is to be noted that she always favored training women and trained 45 women researchers directly under her from 1906 until her death in 1934 at the age of 67. She finally succumbed to health problems due to overexposure to radiation throughout her career.

May I summarize her life by interpreting her words: In life there is nothing to fear—everything can be understood. *Dans la vie, rien n'est à craindre, tout est à comprendre*. Marie Curie's material and emotional difficulties motivated her to be curious enough to successfully create. Her strong family bonds allowed her to survive any difficulty by tedious, tenacious hard work and research. Today, she and her husband rest side by side in a shared tomb at the prestigious Pantheon in the heart of Paris. Her creativity was her work culture of survival. Her life was a stable straight line of creativity.

Gnostic Geniuses Next...

Tajna Nikole Tesle was born in 1856 as one of the five children in the Serbian village of Smiljan, Croatia, under the military rule of the Austrian–Hungarian Habsburg Empire. His mother could not read or write but could recite by heart Serbian poetry by just listening and was very ingenious and dexterous inventing her own homemade household contraptions. His father was a very rigorous and strict person devoting energy and time to mystical religious pursuits. Both Nikola's father and grandfather were Orthodox popes and his mother the daughter of an Orthodox priest. The family name means "hermit" which I allow myself to define, in relation to the life story that follows, as eccentric celibates living in isolation pursuing their own gnostic concepts and ideas or a form of fanatic mystic narcissistic curiosity. Nikola did not respect his father's wishes to pursue the family patriarchal profession as a member of the Orthodox clergy and began his studies in 1875 at the Polytechnic Institute of Graz, Austria, on a local scholarship from military authorities eliminating his material worries and difficulties so typical of many struggling students. He excelled and completed the program of the first 2 years in just 1 year including mathematics, physics, and mechanical engineering but lost his scholarship and stopped studying after the first semester of the third year.

A brilliant student since childhood, he had an exceptional eidetic, or photographic, memory and a unique visual ability never needing models or schemas to learn concepts and principles. After Graz, he worked as an assistant engineer for a year cutting off all contact with family which seems to have been a destabilizing phantom out of the past haunting him the rest of his life. In 1878, his devoted and devout father convinced him to enroll at the University of Charles in Prague where Nikola was greatly impressed and influenced by the Austrian physicist, Ernst Mach. The death of his father meant another interruption in his studies due to financial difficulty, and apparently, he only completed one semester there. It was in 1881 when he finally found work as an engineer in the central government telegraph center of Budapest where he excelled quickly attracting the attention of his superiors which became the basic behavior pattern of his employee identity throughout his professional life. After collaborating with another young colleague on a turbine project for continuous power supply of electricity, he became head electrician before becoming head engineer of the first telephone network of Hungary. He is attributed by some experts to have invented a sort of prototype at the time which later was known as the loud speaker. At this moment in his life, he developed a keen insatiable taste for Hinduism and Sanskrit feeding an innate inherited desire of mystic curiosity manifesting itself over and over again. Thus, always misunderstood by his entourage, it made him look like a lunatic visionary genius. It is important to note here that in Hinduism when the physical senses of the human body are operating, one leaves the superior higher reality and enters the lower material non-reality.

Recruited by Continental Edison Company in 1882, he again excelled and was quickly recognized by his superiors and put in direct contact with Thomas Edison himself. At this time, Nikola invented the first rotating induction motor which was later patented under Edison's name. Nikola naively accepted the personal invitation to come work in America with this very sly and extremely ambitious American inventor and businessman, Thomas Edison. At the tender age of 28, our idealistic unrealistic Serbian immigrant arrives in America to participate in Edison's project of the first electrical distribution network in New York City. He proposes alternative electrical current as a solution to frequent breakdowns and problems, but Edison preferred his own continuous electrical current approach. From this moment on, these two stubborn narcissistic inventors began a lifelong conflict of professional competition. Edison finally agreed and promised a $50,000 bonus to Tesla, but once the alternative electrical current approach was successfully functioning, Edison barely increased Tesla's weekly salary from $10 to $18. Insulted and furious, Nikola quit. In 1886, he created the Tesla Electric Light & Manufacturing Company, but his partners and investors, who controlled his patents, wanted him to invent an arc lamp based on continuous electrical current. Insulted and furious, he quit. Such a delirious fever of being isolated by the curiosity and beauty of his own ideas became a habit and a part of his traditional behavior pattern throughout his career. A year later, Western Union financed the creation of his second company, the Nikola Tesla Company in New York guaranteeing Tesla 50 % of all revenues from patents. Tesla finally fulfilled his dream and invented an alternative current generator; he presented it at the American Institute of Electrical Engineers in 1888. The same year he attracted the attention of another master–mentor, George Westinghouse, who proved to be a lifelong and more just and honest master for our vulnerable Slavic discoverer. Tesla actually preferred to be called a "discoverer."

Tesla's licensing agreement with Westinghouse gave him, at the time, a very generous $2000 monthly salary as a consultant yet provoked the legendary "War of Electrical Currents" between George Westinghouse and Thomas Edison. Both vied to obtain the government contract to set up and supply, on a nationwide scale, electricity in America. Edison attempted to slander the names of Westinghouse and Tesla thus diminishing the credibility of alternative current in favor of his own continuous current approach by electrocuting poor animals in public to show the danger of alternative current. In 1893, Westinghouse and Tesla's alternative current generators won the commercial industrial electricity war against Edison. But World War I came and almost bankrupted both Edison and Westinghouse, and the latter bought out the rights to Tesla's patents for $216,000. Fortunately, for future generations, Tesla's feverish delirium of allowing his curiosity to become his own isolated hermetic world of ideas went on, and he continued to create. Yes, he is the father of electromagnetism but much much more. He remains a legend of creativity because he left behind many inventions that are still used in our world of advanced technologies. He invented what is known as Tesla's coil still used in telecommunications industry today as well as discovered and then published his research about the principle of radar used as a basis for MIT to develop what we know today as the radar system. Tesla's US Patent #613809 is the basis today for robotic remote control, and he is considered by many as the great-grandfather of the "Wi-Fi." The scientific symbol "T," which is the unit of measurement for magnetic flux density, is in honor of him and his work. Larry Page, cofounder of Google, paid Tesla a tribute as the great-grandfather of the Internet. Tesla's transatlantic tower for world global communications and worldwide electrical distribution network, for free electricity bringing world peace, failed, and his Italian rival got the Nobel Prize instead of him. His last patent in 1928 was for a biplane with vertical takeoff and landing properties. In the fall of 1937, he was accidently hit by a taxi as he was walking, as usual, to feed the birds in Central Park but refused all charity, medical care, or hospitalization and never completely recovered. Penniless and alone, in his room at the hotel in New York living off a $125 weekly stipend as a Westinghouse researcher, Tesla passed away at 86. No, Nikola Tesla was not any good at financing or R&D budgets or marketing and industrial commercialization transforming his inventions into moneymaking innovations, but he was, and still is, considered to be the most prolific creator who invented and innovated in the twentieth century. In the recent film, "The Prestige," David Bowie portrayed him as a deep sensitive refined gentleman who admitted he was a slave to American industrial masters. Tesla was a serious scientist and a credible creator whose immense contribution to humanity is still not yet clearly understood. The complex dichotomy of his existence is visually symbolized by the Hungarian bill of money issued in his memory because it is not bankable on the market. Nikola Tesla's feverish almost mystical curiosity and the haunting ever-present phantom of the past—his inherent Slavic servant–master syndrome—became his work culture. His life was a feverish spiral of creativity.

Last but Not Least…

Steve Paul Jobs was born, abandoned, but adopted in 1955. His biological father was a Syrian Muslim and his biological mother a Swiss–American Roman Catholic. They met as students at the University of Wisconsin. He later referred to them as a good sperm bank. His biological parents married 10 months after his birth giving birth to his biological sister, Mona, before divorcing. When searching as an adult for his biological parents, he was so disappointed with what he found out about his father that he never contacted him. His adopted father, a laser machine technician, spent a lot of quality time with him in the family garage taking apart and putting back together electronic devices, machinery, etc., thus giving Steve that precious self-confidence money cannot buy and which most of us spend a lifetime trying to develop. He was a prankster troublesome child paid to do his school work by his fourth grade teacher because he was bored. His parents refused his primary school's suggestion to put him directly into high school. He threatened to never return to school again unless his parents agreed to change him to middle school so they moved just a couple of miles from Mountain View to Palo Alto, California.

In his high school days, a mutual friend introduced him to his would-be future professional soul mate, Steve Wozniak, a student at Berkeley. During those nostalgic high school days, he was sponsored by a neighbor, a Hewlett–Packard computer engineer, into the Hewlett–Packard club for young people interested in computers which was held every Tuesday night in the Hewlett–Packard company cafeteria. One of the guest speakers actually took the group of young people into his Hewlett–Packard lab where Steve saw his first computer, an HP9100A. At the age of 13, he actually telephoned for 20 min to William Hewlett himself, one of the founders of Hewlett–Packard, and asked for parts needed to build a frequency meter. He not only got the needed parts but a summer job at Hewlett–Packard as an assembly line technician the summer after his freshman year in high school. In 1971, the two Steves took seriously an article from *Esquire* magazine about how to build their own "blue box" to tap in on AT&T telephone systems and make free long-distance calls which they also sold. Here are found the undeniable seeds taking root for a lifelong collaboration.

After high school graduation, Jobs preferred to audit the more creative classes as a university-level student like calligraphy at Reed College in Oregon which he says permitted him later to conceive a very rich computer typography. As so many of his generation, Jobs succumbed to hippie tendencies including being vegetarian, gurus, ashrams, LSD, the Beatles and Bob Dylan, etc. Never having completed his university-level degree at Reed, he lived at home and found work as a technician with Atari where he met the second of his future Apple cofounders, Ronald Wayne. He had to work the night shift because colleagues complained about his hippie-inspired no-bathing behavior and smell. After several months traveling around India, he came back home, shaved his head, and wore traditional Hare Krishna style robes. Again at Atari, Jobs asked Wozniak's help to reduce the number of microchips necessary for the game "Breakout" down to a mere 46 microchips but secretly only shared $700 of his $5000 bonus with Wozniak. Profits were already an obsession. Moneymaking marketing simply was an innate skill for our future ICT tycoon. Again, his aggressive American businessman style surfaced very young when he and Wozniak belonged to the Homebrew Computer Club. It was a computer club where members participated, openly exchanged, and freely shared the latest computer developments between each other. At this time, Wozniak, inspired by the Altair computer equipped with Intel 8080, developed a microprocessor later to be used to conceive Apple I. Jobs did not want to openly and freely share the plans and drawings with the other club members but convinced Wozniak to create a company to make and sell them. Jobs sold his "hippie" Volkswagen bus and Wozniak his HP-65 calculator to create Apple in 1976.

After a California business angel invested $250,000 and imposed a business plan, the two Steves built 50 Apple I computers in the family Palo Alto garage selling them to a nearby computer shop. Later a corporation in 1977, they built and sold their Apple II computer increasing company sales by 700 %. As a new public corporation, Apple made $1.2 million at the end of the first day of being quoted on the stock exchange. A multimillionaire at 25, he refused to give a piece of the cake to one of his employees who was one of his oldest friends and loyal workers from the Reed/India hippie days. Steve Jobs, the ICT tycoon, was born. He invited Xerox to invest in Apple to have access to their screen graphics technology. He later quoted Picasso as having said that good artists copy, but great artists steal. Steve Jobs, the ICT tycoon, moved further onward to his destiny. His exclusive team of six young brilliant engineers calling themselves the "Pirates" invented the one-button mouse floating on an inside little rubber ball capable of sliding on a Formica desktop as well as the leg of a pair of jeans. Jobs recruited the president of Pepsi-Cola and, together with Ridley Scott, imagined a television publicity spot for the American football Super Bowl Games in 1984 that revolutionized corporate publicity campaigning, focusing more on the logo rather than on the products themselves.

More than ever before, the phenomenon of Groupthink continued to plague Apple because Jobs was capricious, erratic, unstable, and unpredictable. Steve Jobs' distorted field of reality became his own world of ideas that he wanted the entire world to accept. He became famous among partners, colleagues, and employees for his legendary binary work relationship: "that's great!" or "that's shit!" The board of directors confined him to an isolated office away from the key creators with the title "global-thinking manager." Insulted and hurt, he quit founding a new company called NeXT and invented the interpersonal computer revolutionizing company group and teamwork. Later, he admitted leaving Apple freed him for the most creative period of his career. Jobs bought the computer graphics division of Lucasfilm to create Pixar revolutionizing the world of animated films with his Oscar-winning animated film "Toy Story." After some Disney contracts, he eventually sold his Pixar to Disney for $75 million. When Apple then bought NeXT for $429 million, it put Jobs once again at the top of Apple. "Think differently" was his new commandment, and the new Mac OS X came out on the market. The ICT tycoon, Jobs, became even more narcissist and egocentric. What follows defied the expectations of any and all: iMac, iTunes, iTunes Stores, Apple Stores, iPhone, iCloud, iPod, and iPad. Jobs now claimed his only title as iCEO of the "Real Artist Ship." A charismatic seductive speaker, he made speaking tours in his jeans. He said what he thought and did what he said. This very demanding perfectionist was loved and/or hated everywhere by everyone. His vertical product management proved profitable. A portion of the revenue generated by iPods went to his competitor and friend, Bill Gate's world humanitarian fund, "The Giving Pledge." In a talk to students at Stanford in 2005, he told them: "…follow your heart, your instincts… be crazy." It was actually during a Stanford talk he met his future wife, Laurene. They were married by a Zen Buddhist priest in 1997 and had two children: a son, Erin, and a daughter, Eve. The story is told how he ran across a parking lot to ask her out to dinner, and they never parted. Another daughter, Lisa, was the fruit of his affair in early Apple days with a former girlfriend, Chris Anne, who had struggled to raise Lisa as a single unwed mother before Jobs recognized legally Lisa as his daughter—even though he had already named a computer after her. Chris Anne had believed Steve's excuse of being sterile therefore incapable of being Lisa's biological father. Emotions were never his priority. After a long 7-year battle of health problems and cancer, he passed away in 2011 at 56, and his last words were "Oh wow. Oh wow. Oh wow." He worked up until the night just before he died. His insatiable curiosity to create was amply rewarded by his profit-making work culture; his life was a worldwide whirlwind of creativity.

To End with…

Allow me now to succinctly summarize selective points in common between these exceptionally creative people as a non-exhaustive, hence indicative not representative, comparative analysis within the following framework of criteria considered. Confirmation for some and perhaps shocking for others, only 33 % (Curie) had an outstanding academic background. Interestingly enough, even though 66 % (Curie and Tesla) were of the same generation and from the same part of the world, yet they have the least in common. Outstanding academic accomplishments, age, and origins are not the most important common denominators determining employee capacity for professional creativity. But 100 % share the common denominators (listed below in *italics*) of exceptional intellectual capacities plus being hard working and tenacious employees.

Marie Curie/Nikola Tesla

- Eastern European from large families of five children
- Material and emotional struggles during childhood and youth and in a new country
- Lacked financial and marketing skills
- *Exceptional intellectual capacities*
- *Hard working and tenacious employees*
- Improved the world we live in

Nikola Tesla/Steve Jobs

- American engineers who never completed university
- Isolated in their own world of ideas
- Founded and lost companies
- Key investors at key moments
- Victims of internal political power games, quitting jobs out of anger
- Unstable explosive complex personalities making work relationships very difficult
- Adhered to alternative mystical and spiritual thinking
- *Exceptional intellectual capacities*
- *Hard working and tenacious employees*
- Revolutionized the world we live in

Steve Jobs/Marie Curie

- Married with close strong marital relationships and had their own families and children.
- Their fathers frequently spent quality learning time with their children.
- Severe health problems including fatal diseases (cancers).
- *Exceptional intellectual capacities.*
- *Hard working and tenacious employees.*
- Improved the world we live in.

In conclusion, I propose the following deductions about these three highly creative people.

The professional life and resulting creativity of Marie Curie was greatly influenced by a practical Polish curiosity cultivating *a survival-oriented work culture.*

The professional life and resulting creativity of Nikola Tesla was highly influenced by a mystic Slavic curiosity cultivating *a servant-oriented work culture.*

The professional life and resulting creativity of Steve Jobs was definitely influenced by an ambitious American post-hippie "be anything and do everything" curiosity cultivating a *profit-oriented work culture.*

To end with, here is a little down-to-earth common sense from a wise Canadian, Henry Mintzberg, who claims the best of management methods is the one that works. Let the same be said about being creative. Professional creativity is a developed work culture transcending any confining mother cultural constraints. It is simply being curious enough to cultivate your own way of seeing and doing as a devoted wife and mother or as a misunderstood immigrant bachelor or as a wealthy hippie tycoon. Creativity is a work culture of seeing and doing in a different way.

Chapter 12
Cultural Underpinnings in Entrepreneurship

Igor N. Dubina and Suzanna J. Ramos

Abstract The issue of what cultural aspects impact entrepreneurial behavior is examined empirically most often on the basis of Hofstede's (*Culture's Consequences: International Differences in Work-Related Values*, 1980) model. Hofstede's research is useful to understand behaviors and attitudes at work, such as leadership, motivation, or the behavior and relationships between members and how these factors affect the level of entrepreneurship in any given society. Hofstede's model of cultural dimensions serves as the basis for theoretical and empirical research over the past 30 years on how national culture influences business and management. This chapter is a description of Hofstede's canonical four cultural dimensions represented in his model and their relationships with entrepreneurship.

Introduction

To define national culture, researchers and practitioners generally use national boundaries as a proxy for differentiating between distinct cultures and their respective cultural characteristics. Similarities in national cultures derive from common history, religion, geography, and language. Although there are differences within national borders, these differences are assumed to be of less significance than those found between nations. The concept of national culture refers to a set of values, beliefs, and attitudes shared by individuals of a human group, which in turn influences individuals' behavior and social relationships (Hofstede 1980, 2001). The interactive aggregate of common characteristics that influence a group's response to its environment is what distinguishes one group from another.

I.N. Dubina, Ph.D., Dr.Sc. (✉)
Novosibirsk National Research State University, Novosibirsk, Russia

Altai State University, Barnaul, Russia
e-mail: igor_dubina@yahoo.com

S.J. Ramos, Ph.D.
Department of Educational Psychology, College of Education and Human Development, Texas A&M University, 3704B Plainsman Lane, College Station, TX 77843, USA
e-mail: suzannaramos@gmail.com

I.N. Dubina, E.G. Carayannis (eds.), *Creativity, Innovation, and Entrepreneurship Across Cultures*, Innovation, Technology, and Knowledge Management, DOI 10.1007/978-1-4939-3261-0_12

The main driving force of national culture is its values. Cultural values determine which behavior is appropriate, thus influencing an individual's choice. The result is persistent differences in human behavior in different countries. Consequently, when people establish organizations, the characteristics of these organizations reflect the people's cultural values. Against this backdrop, cultural factors exert an influence on entrepreneurship—a process that involves identifying an opportunity, developing a business concept, acquiring the necessary resources to implement the concept, and then managing the venture to reap benefits.

Numerous studies have linked national culture to the strategic decision-making process that occurs within entrepreneurial organizations, particularly the cultural dimensions that contribute to entrepreneurship. This endeavor has highlighted the importance of sociocultural variables in explaining variations in entrepreneurship and economic development. Research interest has focused on understanding the influence of national cultural values on the individual entrepreneur, as well as the cultural variables in explaining variations in entrepreneurial motivation, orientation, and behavior across countries. If different cultures possess different attitudes toward entrepreneurship, then it follows that certain cultures engender more entrepreneurial behavior than others.

The issue of what cultural aspects impact entrepreneurial behavior is examined empirically most often on the basis of Hofstede's model (1980). Hofstede contends that a nation is a social system which has its own culture—legal, educational, and political systems—which serves as a vehicle for mental programming of the nation's members. Therefore, a nation possesses its own culture. Hofstede's (1980, 2001) study on national cultures within organizations not only analyzes national cultures, but also demonstrates the effects of cultural differences inside organizations. His research is particularly useful to understand behaviors and attitudes at work, such as leadership, motivation, or the behavior and relationships between members and how these factors affect the level of entrepreneurship in any given society.

Hofstede's model of cultural dimensions serves as the basis for theoretical and empirical research over the past 30 years on how national culture influences business and management (Hofstede 1980, 1993, 2001). The first four dimensions of the model were initially detected through a comparison of the values of similar people (more than 100,000 employees and managers) in 64 national subsidiaries of the IBM Corporation (Hofstede 1980). People working for the same multinational company, but in different countries, represent well-matched samples from the populations of their countries, similar in all respects except nationality. The following is a description of the four cultural dimensions represented in the model and their relationships with entrepreneurship.

Power Distance

Power distance describes the degree "to which less powerful members of institutions and organizations within a country expect and accept that power is distributed unequally" (Hofstede 2001, p. 98). This dimension measures how far inequality is accepted by a culture.

Cultures with a high score in power distance show existential inequality between the superior and the subordinate. A high-power distant organization will most likely use formal forms of address or those that reflect status differences. Such organizations centralize power in a few hands as much as possible. Inequality is an inherent feature of society, so it accepts and expects more powerful individuals to possess certain privileges. High-power distance countries include China with a power distance index (PDI) of 90, Russia (PDI: 93), and Malaysia (PDI: 104) (Hofstede 2001).

In contrast, low-power distance cultures accept that power be distributed equally. Equality is valued, and those with more power or status should not act in ways that call attention to their advantages. Hierarchies are not as firmly established compared to high-power distance cultures. Cultures with a low-power distance index include Austria (PDI: 11), Norway (PDI: 31), and the USA (PDI: 40) (Hofstede 2001).

Mitchell et al. (2000) contend that high power distance has a negative effect on venture creation. Their argument is based on the fact that in such societies, individuals from the lower classes may consider firm creation to be restricted to the elites—individuals who have access to both the necessary resources and experience. Consequently, the majority of the population outside this small group of elites will fail to develop valid cognitive schemas either for evaluating environmental opportunities or for undertaking entrepreneurial ventures. Further, since entrepreneurs have high needs for achievement and independence, Hofstede (1980) found that power distance negatively correlated with a belief in the importance of independence.

Uncertainty Avoidance

Uncertainty avoidance refers to "the extent to which the members of a culture feel threatened by uncertain or unknown situations" (Hofstede 2001, p. 161). This dimension measures the extent to which members of a culture fear uncertain or unknown situations, reflecting the society's intolerance of ambiguity and uncertainty.

Strong uncertainty avoiding countries typically feel threatened by ambiguous situations and design ways to reduce their stress and fear of the unknown. The stress resulting from uncertainty leads to the need to adopt formal rules and to be less tolerant of people, groups, or ideas that diverge from already familiar models. Individuals fear failure in these societies, and consequently, their members avoid undertaking risks. Examples of countries with a relatively high-uncertainty avoidance index (UAI) include Argentina (UAI: 86), Turkey (UAI: 85), and Italy (UAI: 75) (Hofstede 2001).

Low-uncertainty avoidance societies fully accept uncertainty. Such countries exhibit a higher level of tolerance for change and ambiguity and accept and often embrace the risks associated with an uncertain future. In societies with low uncertainty avoidance, organizational rules can be violated for pragmatic reasons. Conflicts and ambiguous situations constitute a natural part of life in an organization. The USA (UAI: 46), Malaysia (UAI: 36), and South Africa (UAI: 49) are examples of low-uncertainty avoidance societies (Hofstede 2001).

Low-uncertainty avoidance cultures are interested in exploring new ways of doing things, although this necessarily implies some level of uncertainty. Individuals in these cultures are more ready to assume risks and exploit any opportunities they can identify in their environment, even in situations where information is limited (Busenitz and Lau 1996). All these behaviors create an atmosphere where the members of these cultures are inclined to manifest more entrepreneurial inclinations.

Individualism Versus Collectivism

This dimension of culture describes "the relationship between the individual and the collectivity which prevails in a given society" (Hofstede 2001, p. 209). The individualism–collectivism dimension shows whether the interests of an individual or a group are more important. According to this dimension, all cultures can be characterized by the strength of social forces, which bring individuals together to form social entities.

Individualism is the degree to which people in a country prefer to act as individuals rather than as members of groups. Individualistic societies are characterized by an emphasis on individual initiative, self-sufficiency and control, and the pursuit of individual goals that may or may not be consistent with in-group goals or achievement. In an individualistic environment, people are motivated by self-interest and achievement of personal goals. They are hesitant to contribute to collective action unless their own efforts are recognized, preferring instead to benefit from the efforts of others. Examples are the USA (IDV: 91), Canada (IDV: 80), and New Zealand (IDV: 79).

On the contrary, people in collectivistic societies are connected to each other through strong and cohesive groups that protect them during their lives. It is assumed that people are loyal to these groups. In collectivistic cultures, there is a communal-based regulation of society. People connect their identity with groups more than with other characteristics of personality. Collectivism involves the subordination of personal interests to the goals of the larger work group; an emphasis on sharing, cooperation, and group harmony; a concern with group welfare; and hostility toward out-group members. Collectivists believe that they are an indispensable part of the group and will readily contribute without concern for advantage being taken of them or for whether others are doing their part. They feel personally responsible for the group product and are oriented toward sharing group rewards. Countries with a relatively low index of individualism include China (IDV: 20), the Philippines (IDV: 32), and Indonesia (IDV: 14).

Recent research on entrepreneurship has mostly focused on the issue of the role of individualism and collectivism. Individualism–collectivism seems to be one of the more salient dimensions of culture insofar as entrepreneurship is concerned. It is suggested that of all the elements necessary for successful entrepreneurship, the independent entrepreneur is the most critical. Individual autonomy and a sense of ownership of innovation encourage the risk-taking and persistence required for entrepreneurship.

Therefore, in individualistic cultures, entrepreneurship is valued and encouraged by the society since the entrepreneurial individuals usually show high levels of personal confidence, initiative, and courage. For example, individualism is an intrinsic aspect of American culture, which helps to explain the relatively intensive amount of independent entrepreneurial activity in the USA.

In collectivist societies, both private property and the protection of individuals' ideas are limited. In these societies, in which collective economic activity predominates, there may be fewer opportunities for individuals to develop the capabilities and skills needed to create new firms. However, it should also be pointed out that in the context of Asian cultures that are more collective and relationship oriented than Western cultures, entrepreneurship may be more of a family or group emphasis than on individual endeavor.

Masculinity and Femininity

This dimension reflects a culture's dominant values with respect to achievement, recognition, competitiveness, and interpersonal relationships (Hofstede 2001). Masculinity–femininity shows to what extent a culture is dominated by such masculine values as orientation toward achievement and competition. The detection of self-assertiveness and other "masculine" values, such as independence and career, refers to masculinity, while discretion, tolerance, and solidarity describe feminine behavior. Hofstede (2001) describes a masculine culture as one in which "people live to work" (longer work hours and short vacations) and a feminine culture as one in which "people work to live" (longer vacations and flexible working hours).

Masculinity as one pole of this cultural dimension is highlighted in societies where the social gender roles are clear: men are supposed to be assertive, tough, and oriented toward material success. Masculine societies are aggressive and task and performance oriented, with many occupations typically considered gender specific. In a masculine society, challenge, advancement, and the accumulation of money are important. Countries with a relatively high MAS score are Japan (MAS: 95), Italy (MAS: 70), and Ireland (MAS: 68) (Hofstede 2001).

Femininity, on the other hand, characterizes societies in which the social gender roles overlap: both men and women are assumed to be modest, sensitive, and concerned about the quality of life. Feminine societies are characterized by an emphasis on relationships and social interactions. Economic growth may not necessarily be the primary concern of the society. In a feminine culture, a friendly atmosphere, job security, and cooperation are paramount. Such countries include Spain (MAS: 42), Thailand (MAS: 34), and Portugal (MAS: 31) (Hofstede 2001).

Masculinity of a society refers to assertiveness, competitiveness, and achievements. The Western concept of the entrepreneur follows the "hero" metaphor. It is argued that without the visionary leadership and persistence demonstrated by this individual, little will be accomplished. In masculine countries, individuals are taught to appreciate strong and independent heroes who personify superiority

(Steensma et al. 2000a, b). These cultures view failure as a sign of mediocrity, so the need for achievement intensifies. Conversely, individuals from feminine countries tend to be less aggressive and assertive as they are taught by their societies to seek mutual gain (Hofstede 2001). With these arguments, masculine cultures are associated with more entrepreneurial behavior—the higher the masculinity level in a particular area, the higher the level of entrepreneurial behavior.

Conclusion

Generally, cultural values identified with the classical Hofstede's four-dimensional model and associated with entrepreneurship are high power distance, high individualism, low uncertainty avoidance, and high masculinity (McGrath et al. 1992; Swierczek and Quang 2004). Countries with these features are more entrepreneurial (entrepreneurship rates defined as the percentage of new business owners in a country). The classical model of the four cultural dimensions has been further developed for the last three decades. A fifth dimension, long-term orientation (LTO), was added in 1991 based on research by Michael Bond (Hofstede 2001) and modified later (Hofstede and Minkov 2010). Hofstede et al. (2010) then added a sixth dimension, indulgence versus restraint. There are also several other models of cultural dimensions that have been suggested in contemporary publications, e.g., Schwartz and Sagiv (1995), Trompenaars and Hampden-Turner (1998), and House (2002). However, within a global context of entrepreneurship, there are still contradictive findings published in the literature and a limited understanding of the extent of a particular culture's influence on entrepreneurship and entrepreneurial behavior. There is a need for further examination of the relationship between cultural characteristics and individual and organizational opportunity-seeking propensities and the peculiarities of venture-creation decisions across cultures.

References

Busenitz, L. W., & Lau, C. M. (1996). A cross-cultural cognitive model of new venture creation. *Entrepreneurship: Theory and Practice, 20*(4), 25–39.

Hofstede, G. (1980). *Culture's consequences: International differences in work-related values*. Beverly Hills, CA: Sage Publications.

Hofstede, G. (1993). Cultural constraints in management theories. *Academy of Management Executive, 7*(1), 81–94.

Hofstede, G. (2001). *Culture's consequences: Comparing values, behaviors, institutions, and organizations across nations*. Thousand Oaks, CA: Sage Publications.

Hofstede, G., Hofstede, G. J., & Minkov, M. (2010). *Culture and organizations - Software of the mind: Intercultural cooperation and its importance*. New York: McGraw-Hill.

Hofstede, G., & Minkov, M. (2010). Long- versus short-term orientation: new perspectives. *Asia Pacific Business Review, 16*(4), 493–504.

House, R. (2002). Understanding cultures and implicit leadership theories across the globe: an introduction to project GLOBE. *Journal of World Business, 37*(1), 3–10.
McGrath, R. G., MacMillan, I. C., & Scheinberg, S. (1992). Elitists, risk-takers, and rugged individualists? An exploratory analysis of cultural differences between entrepreneurs and non-Entrepreneurs. *Journal of Business Venturing, 7*, 115–135.
Mitchell, R. K., Smith, B., Seawright, K. W., & Morse, E. A. (2000). Cross-cultural cognitions and the venture creation decision. *Academy of Management Journal, 43*(5), 974–993.
Schwartz, S. H., & Sagiv, L. (1995). Identifying culture-specifics in the content and structure of values. *Journal of Cross-Cultural Psychology, 26*(1), 92–116.
Steensma, H. K., Marino, L., Weaver, K. M., & Dickson, P. H. (2000a). The influence of national culture on the formation of technology alliances by entrepreneurial firms. *Academy of Management Journal, 43*, 951–973.
Steensma, H. K., Marino, L., & Weaver, K. M. (2000b). Attitudes toward cooperative strategies: A cross-cultural analysis of entrepreneurs. *Journal of International Business Studies, 31*, 591–609.
Swierczek, F. W., & Quang, T. (2004). Entrepreneurial cultures in Asia: business policy or cultural imperative. *Journal of Enterprising Culture, 12*(2), 127–145.
Trompenaars, F., & Hampden-Turner, C. (1998). *Riding the waves of culture: Understanding cultural diversity in global business* (2nd ed.). New York: McGraw Hill.

Afterword

As we endeavored to demonstrate in this monograph, researchers and practitioners who deal with business, creativity, and innovation across societies should investigate culture. As Franke et al. (1991) noted, "with business becoming more and more international, profiles of national culture can become tools for strategic choices in corporate boardrooms. Sensitivity to cultural variables will be needed for decisions as to what to do in which countries… Our findings suggest that we should view human values as serious business." Indeed, studies over the past several decades in the area of national culture and values, and more recent studies (including those presented in this monograph) concerning the effect of culture on innovation, support two basic conclusions: (1) national differences in culture and values exist, and (2) certain cultural characteristics can support or block creative, innovative, and entrepreneurial activities more than others. As we see, certain cultural profiles can be more conducive for creativity and more "innovation-friendly" than others and better suited than others to support creativity, innovation, and entrepreneurship.

Among existing research, the chapters included in this book indicate that culture indeed plays a part in creativity and innovation processes. Cultural impact on creative, innovative, and entrepreneurial activity may be put into effect through indirect pressure on cognitive and behavioral patterns, personality, motivation, etc., or through direct effect of the "desired" activity, available opportunities in certain domains, or role models.

Although a research link between national culture and innovation becomes better decade to decade in terms of strengthening theoretical and empirical foundation, it is still obvious that more work should examine the interaction of culture and phenomena of creativity, innovation, and entrepreneurship. In spite of active and wide application of different models of cultural dimensions, as well as the sizable corpus of research findings, the consequences of the cultural differences for creativity, invention, innovation, and entrepreneurship are still far from clear.

First of all, more empirical research attempting to tie innovation and entrepreneurship to cultural dimensions are needed. Secondly, the preliminary models and theories

I.N. Dubina, E.G. Carayannis (eds.), *Creativity, Innovation, and Entrepreneurship Across Cultures*, Innovation, Technology, and Knowledge Management, DOI 10.1007/978-1-4939-3261-0 require considerable testing in order to provide the basis for future conceptual and empirical research on this important topic. Thirdly, there is still a need to better consolidate the findings, particularly in support of developing a genera theoretical framework of a cultural impact in creativity, innovation, and entrepreneurship.

In this concluding section, I would like to specially note a dominant paradigm and position of Hofstede's model in recent cross-cultural studies of innovation and entrepreneurship. Such a position may constrain in some sense further development of cross-cultural studies in this field. Although Hofstede's framework has continuously been developing, a significant number of studies, including those we mentioned in this monograph, demonstrate that Hofstede's dimensions are not enough for explanation of cultural impacts. Hofstede's model is undoubtedly helpful, but it has already become a paradigm of the "normal" cross-cultural science in terms of Thomas Kuhn. So, proliferation of approaches (somewhat a "methodological anarchism" as Paul Feyerabend said) in this field is needed. Further research should increase the number of cultural variables employed to validly and reliably explain cross-cultural variations of creativity, innovation, and entrepreneurship, as well as variables validly and reliably characterizing these very phenomena themselves.

Clearly, culture is only one of many factors, which may multilaterally influence creativity, innovation, and entrepreneurship. Moreover, cultural influence is undoubtedly concealed by many other factors (economic, social, historical, geographical, political, etc.). As Dr. Black, one of this book's respectful authors, observed during his numerous travels over the world, cultural influence often not so clearly appears because of globalization and social, cultural, and historical dynamism. This idea was also expressed by Wu Guanzhong, a contemporary Chinese artist: "Between the Chinese and foreigner of today there exists a distance, but far greater distance exists between the Chinese of today and the Chinese of antiquity. While the former distance will diminish with time the latter distance will lengthen" (cited in Rudowicz, p. 79).

Concluding this book, I would like to sincerely and deeply thank again all of the authors of this monograph's chapters, as well as Prof. Elias Carayannis, this Book Series Editor, Nicholas Philipson, the Editorial Director of Springer Science+Business Media, and all the Springer team who worked in order to bring this book to reality. I hope that readers of this book will also appreciate my colleagues' efforts they put into this monograph and their valuable contribution to understanding and explanation of multilateral issues and variances of creativity, innovation, and entrepreneurship across cultures.

Sincerely,
Igor N. Dubina, the Book Editor

References

1. Franke, R. H., Hofstede, G., & Bond, M. H. (1991) Cultural roots of economic performance. *Strategic Management Journal, 12*(S1), 165–173.

Index

A
Achaemenid period, 78
Achievement, 3, 14, 16, 19–21
Adaptive style of creativity
  innovative style of creativity, 59
  mean rating, 55, 56
Adaptor
  characteristics, 54
  implicit perceptions, 56
  Kirton's contention, 53
  minimum and maximum ratings, 55
Alexandria, 80, 83
Amabile, T.M., 63, 125
Amado, G., 2, 11
American managerial model, 12
Anatolia, 78
Ando, T., 73
Antiochus IV Epiphanes, 82
Are You a Crayon Breaker™, 37
Arieti, S., 30
Arsacid, 78, 82–85
Asian cultures, 13
Assertiveness, 14, 18
Aycan, Z., 4

B
Balkan, 78
Barsoux, J.-L, 127
Behavior, 1–21
Benedict, R., 2
Benevolence, 20, 21
Black, R.A., 35–50
Bond, M., 18
Borumand, S., 77–85
Brain activity, 88
Buddhism, 12
Burns, T., 115
Byeong Yong, K., 124

C
Chimento, M.D., 57
Chinese, 53–56, 59
Chinese individual creativity
  advantages and disadvantages, 72–74
  educational background, 70, 71
  industrial structure, 70
  natural environment, 70
  one-child policy, 72
  political system, 70, 71
  politics and business, 70
  students' performance, 71
  training grow, 72
Collectivist societies, 151
College entrance examination, 71
Communication
  individuals' experience, 96
  and interaction, 98
Conformity, 10, 14, 20
Confucian dynamism, 15, 18
Constellations of meaning, 102
Creative traits
  trusting, natural creative traits, 40–41
Creativeness, 88, 93
Creativity, 30–33, 36–37, 39–49, 51, 63

I.N. Dubina, E.G. Carayannis (eds.), *Creativity, Innovation, and Entrepreneurship Across Cultures*, Innovation, Technology, and Knowledge Management, DOI 10.1007/978-1-4939-3261-0

Creativity (*cont.*)
activity, 64
adaptive style (*see* Adaptive style of creativity)
British Empire, 47
changing cultures, 43
characteristics, 63
cognitive process, 29
complete openness, 49
contrasts and integration of contrasts, 46
creative soul, 45–46
creativogenic society, 30
cultures, 30, 44
definition, 53, 64
explicit theories (*see* Explicit theories of creativity)
group (*see* Group creativity)
implicit theories (*see* Implicit theories of creativity)
individual (*see* Individual creativity)
innovative style (*see* Innovative style of creativity)
inventiveness
breaking/rejecting traditions, 32
characteristics, 33
meta-cognitive and affective skills, 31
co-operation and conformity, 32
cosmological process, 31
cultivating, 33
cultural influence, 33
ethnic groups, 33
mistakes and losing face, 33
observable product, 31
4-P model, 30
principals, 33
product-centered conception, 31
social influence, 33
spiritual/social forces, 31
team-oriented approach, 33
"western" and "eastern" cultures, 31
Istanbul, 45
juxtapositioning in time, 48
learning
changing perspectives, 49
control and orderliness, 46
creative spirit, 47–48
creativeness and creative thinking skills, 42
daily frustrations, 39
experiencing cultures, 41
from East, 45
juxtapositioning imagination, 48
subconscious/intuition, 40
traits, cultural immersion, 42–43
wandering, 41
weather, 39–40
Western creative meets Eastern creativity, 43–44
multi-faceted phenomenon, 29
personal definitions, 52
philosophical, religious and ethical fundamentals, 32
physical environment, 43
process, 3, 64, 97, 126
products, 64
refined control to refined individualism, 47
rejuvenation, 47
social context, 29
socio-cultural factors, 30
Sri Lanka, 44
traits
Are You a Crayon Breaker™, 37
TTCT™ Traits, 36–37
trusting natural creative traits, 40–41
universal meaning, 30
West and East dichotomy, 30
Western World, 46
Creativogenic society, 30
Critical thinking, 104
Cross-cultural challenges
action nets/communal processes, 95
ambiguities, 96
consonance, 102
economic institutions, 102
entrepreneurship, 102
expectations, 104
immigrants, 103
interlingual sharing, 96
knowledge, 96
multiethnic firms, 102
organizational scholars, 96
performative perspective, 103
positive outcomes, 104
Somalia and Romania, 103
workplace dynamics, 103
Cross-cultural differences, 7
Cross-cultural collaboration, 96
Cross-cultural innovation, 96, 98–104
Cross-cultural management, 4, 101–104
Cross-cultural studies, 1, 2, 9, 11, 18
Cross-cultural synergy, 125–127, 131
Csikszentmihalyi, M., 52
Cultural background
competitiveness and innovation, 113
economic growth, 117
knowledge management, 110
performance orientation, 114

social learning, 113
societies, 110
Cultural context, 135
Cultural convergence
Achaemenid period, 78
Alexandria, 80
Antiochus IV Epiphanes, 82
Babylon, 81
the city of Kingship, 82
Epigoni (successors), 81
ethnic group, 80
federalism and tolerance, 83
government of Seleucids, 82
Greco-Persian mutual relationship, 78
Greeks and Macedonians, 79
Hellenistic period, 82
history of globalization, 78
History of Rome, the Syrian Wars, 81
IT and ICT, 77
Kallias (Callias) treaty, 78
The Library of History, 79
The Life of Alexander, 79
Macedonia, 81
Mithradates I, 83, 84
Nicomedia, 81
The Parallel Lives, 79
Persian rule, 78
Persian Wars/Greco-Persian Wars, 78
Seleucid period, 80
Seleucid rulers' territory, 82
Telephanes communicate, 79
Cultural differences, 2, 3, 88, 110, 125, 127, 148
Cultural dimensions, 11, 14–21, 148, 152
achievement values, 21
assertiveness, 18
benevolence values, 21
conformity values, 20
future orientation, 19
gender egalitarianism, 19
GLOBE, 18
Hedonism values, 21
Hofstede's model, 14–18
humane orientation, 19
in human practices, 14
in-group collectivism, 19
institutional collectivism, 19
performance orientation, 18
power distance, 19
power values, 21
security values, 21
self-direction values, 20
stimulation values, 21
taxonomy, 20
tradition values, 20
uncertainty avoidance, 19
universalism values, 20
Cultural diversity, 2, 121–132
Cultural layers/levels, 5
Cultural synergy
community stakeholders, 126
concept, 125
diversity and creativity, 122
globalization, 121
innovating synergy, 126
micro- and macro-levels, 126
multicultural staff, 122
multinational organizations, 126
Cultural traits, 110, 112
Cultural values, 2–4, 8–11, 18
Culture, 1–21
Anglo-Saxon, 59
Asian, 59 (*see also* Creativity)
in creativity studies, 52
Eastern culture, 53, 59
implicit theories of creativity, 53
Latin, 59
national culture, 53
Singapore, 59
USA, 59
Western culture, 53, 59
Curiosity, 40, 42, 135–145

D
Dali, 49
Darwin, 111
Darwinism, 12
Davis, G.A., 52
Davis, H.D., 6
Decision-making process, 148
Del Giudice, M., 95–104
Distributed cognition, 99
Divergent thinking, 43
Druker, P., 115
Dubina, I.N., 1–21, 29–34, 51–60, 147–152

E
Eastern culture, 53, 59, 60
Electromagnetism, 140
Emotions, 90–92
English, 87–89
Entrepreneurial behavior
collectivistic societies, 150
cultural aspects, 148
decision-making process, 148
individualism *vs.* collectivism, 150
masculinity and femininity, 151

Entrepreneurial behavior (*cont.*)
national culture, 148
power distance, 148
private property, 151
researchers and practitioners, 147
uncertainty avoidance, 149, 150
Entrepreneurial risk, 112
Entrepreneurship
advantage, 112
cultural underpinnings, 147–152
and managerial systems, 2
in society, 112
Epigoni (successors), 81
Esaki, M., 87–93
Esquire magazine, 141
Ethnic groups
categories, 58, 59
Chinese, 53
Indians, 53
Malay, 59
Malays, 53
Singapore, 53, 55
Eureka (West), 32
Evolutionary theory, 110–111
company organization, 111
enterprise decisions, R&D spending, 111
principles, 111
Explicit theories of creativity, 52
KAI theory (*see* Kirton's adaption and innovation (KAI) theory)

F
Fayolle, A., 5
Femininity, 151
Fogg, P.T., 35
France, 19, 36, 49, 129, 130, 137
French personalist and social model, 12
French–Russian analysis, 127
French–Russian collaboration, 127
French–Russian Cultural Diversity, 127–129
French–Russian Partnership Synergy, 129–131
Future competitive advantage, 112
Future orientation, 18, 19

G
Gender egalitarianism, 18, 19
Georgsdottir, A., 13
Gestures, 90–92
Giudice, M.D., 95–104
Glaveanu, V.D., 51
Global leadership, 2
GLOBE program, 18–21
Gonzalez, M., 57
Great Recession, Europe (Case study), 116–118
Greco-Persian mutual relationship, 78
Greek, 31, 77–85, 125
Group consciousness, 65–69
Japan (*see* Japanese group creativity)
Group creativity, 65–69
Japan (*see* Japanese group creativity)
Guilford, J.P., 52, 63
Guseva, N.I., 121–132

H
Hampden-Turner, C., 152
Handy, C., 115
Harvard, P.S., 135–146
Hedonism, 14, 20, 21
Hellenes, 81, 82
Hofstede, G., 6, 9, 11, 14, 17, 18, 114, 127, 148, 149, 151
Hofstede's model of cultural dimensions, 148
business and management, 14
characteristics, 14
conceptualization, 17
criticisms, 17
framework, 16
individualism *vs.* collectivism, 15, 16
long-term vs. short-term orientation, 16
masculinity vs. femininity, 16
non-independent structure, 17
power distance, 15
uncertainty avoidance, 15
Holden, N.J., 95–104
Hoppe, M.H., 11
House, R., 6, 114
Humane orientation, 18, 19

I
Imitation, 64
Implicit theories of creativity
characteristics and experiences, 51
Eastern culture, 53
ethnic groups, 53
expression, 52
I-Paradigm, 52
KAI (*see* Kirton's adaption and innovation (KAI) theory)
laypeople, 52, 53
mental prototypes, 52
physical and social milieu, 52
psychometric approaches, 52
Western culture, 53
Indians, 53–56, 58, 59

Individual consciousness. *See* Chinese individual creativity
Individual creativity, 70–72
Chinese (*see* Chinese individual creativity)
environment, 63
Individualism include China (IDV), 150
Individualism *vs.* collectivism, 14–16
Individualistic cultures, 15, 151
Indulgence *vs.* restraint, 16
In-group collectivism, 18, 19
Inkster, I., 12
Innovation
ad hoc translations, 99
communication and interaction, 98
competitive contests, 98
conceptual effort, 99
creativity, 97
cross-cultural, 99
discovery, 97
economic institutions, 97
integral element, 100
interdependencies, 101
inter-firm network, 98
invention, 97
investment, 97
KM community, 100
KM literature, 101
management, 95–104
rewarding, 98
translation, 99
triangle, 100
Innovation culture
and competitiveness
background, 113, 114
entrepreneurial culture, 112
grid of values, society, 111
individualistic societies, 112
individuality/collectivity, 114
organizational structure, 115
performance orientation, 114
social learning, 113
cultural background, societies, 110
entrepreneurial opportunities, 110
evolutionary theory and R&D, 110–111
Great Recession, Europe (case study), 116–118
organizational culture, 110
social capital, 109
Innovative creativity, 96
Innovative style of creativity, 55, 57, 59
Innovators
and adaptor, 53
anthropologists, 135
characteristics, 54
implicit perceptions, 56
Kirton's, 57
Kirton's contention, 53
laypeople's implicit theory of creativity, 59
Malay group, 56
mean ratings, 55–57
minimum and maximum ratings, 55
perceptual bias, 59
style of creativity, 58
t-tests, 56
Institutional collectivism, 18, 19
Institutional theory, 10, 11
International business, 2
Interrelationship, 77–85, 91, 100
Intra-cultural variations, 7
I-Paradigm, 51, 52
Iran, 78, 81, 82, 85
Invention, 3, 9, 13, 64, 68, 69, 96, 97, 99, 140
Inventiveness, 12, 30–34, 77–85

J
Japanese, 88–92
Japanese group creativity
advantages and disadvantages, 72–74
analytical instrument maker, 68
A School Sports Day, 65, 66
biological macromolecules, 68
classroom, 66
group members' contribution, 68
LED, 68
outcomes, 68
relation to the industrial structure, 67
relation to the natural environment, 67
relation to the political system, 67
service inventions, 69
Shuji Nakamura, 69
strength and weakness, 69
Jones, G.K., 6

K
Kagawa, Y., 57
Kirton, M.J., 52, 54, 57
Kirton's adaption and innovation (KAI) theory, 55
adaptor (*see* Adaptor)
explicit theory, 52
innovator (*see* Innovator)
laypeople's implicit views of creativity, 53
materials, 54
method, 53
Kluckhohn, F., 127

Knowledge-creating company, 101
Koshino, J., 73
Kreiser, P.M., 7

L
Language, 6, 52, 54, 71, 78, 82–84, 87–93, 95, 99–103, 147
Laypeople's, 52, 53, 57, 59
Leung, K., 59
Light-emitting diode (LED), 68
Long-term *vs.* short-term orientation, 15, 16
Losing-of-face effect, 13
Lubart, T.I., 13

M
Macedonians, 79–81, 84
Magnetism of metals, 137
Malays, 53–56, 58, 59
Management process, 124, 127, 129
Management strategies, 2
Masculinity, 151
  vs. femininity, 14, 16
Measurement, 4–9
Melvin Tumin's, 80
Mitchell, R.K., 149
Mithradates I, 83
Miyazaki, H., 73
Moneymaking marketing, 142
Moran, R., 126
Multicultural teams, 122, 129
Muneyoshi, H., 57

N
Nakamura, S., 68, 69
National cultures, 2–8, 10, 11, 13–16, 53, 59, 60
Nishihama, C., 87–93
Novelty, 12, 21, 59, 97, 98
Normal science, 18

O
Okuyama, K., 73
Olympic mathematics, 71
Organizational characteristics, 10
Organizational cultures, 5, 6, 11, 115
Organizational structure, 115, 116
Outward development, 74
Ozawa, S., 73

P
Parthian, 78, 82, 83
Parthian Pahlavi, 83
Peloponnesian War, 78
Peppas, S.C., 7
Perceptual bias, 57, 59
Performance orientation, 18
Persia, 78, 80, 84
Persians, 78, 79, 81, 82, 84, 85
Personal definitions of creativity, 52
Peruta, M.R.D., 95–104
Petrakis, P.E., 109–118
Philip II (359–336 B.C.E.), 79
Phillips, M.E., 4
Picasso, 49
4-P model, 30
Polonium and radium, 138
Power, 11, 15, 19–21
Power distance, 11, 14, 15, 18, 19, 148, 149
Power distance index (PDI), 149
Product-centered creativity, 31
Project team, 74
Puccio, G.J., 57

R
R&D, 110–111
R&D globalization, 2, 3
R&D spending, 111
Radioactivity, 137
Ramos, H., 1–21
Ramos, S.J., 1–21, 29–34, 51–60, 147–152
Ravlin, E.C., 4
Risk-taking, 13, 19, 20
River Halys, 78
Roembke, L, 122
Runco, M.A., 9, 59
Russia, 125, 127–129
Russians fundamental shifts, 131

S
S.P.R.E.A.D.ng™, 35
Sackmann, S.A., 4
Sagiv, L., 152
Satori (East), 32
Schneider, S., 8, 127
Schwartz, S.H., 152
Security, 16, 20, 21
Self-Direction, 14, 20
Service inventions, 69

Shakespeare, W., 47
Shame effect, 13
Shinto conception, 12
Shoguns, 67
Social capital, 110
Social learning theory, 113
Societal attitudes, 10
Socratic and confucian traditions, 32
Stalker, G.M., 115
Stimulation, 20, 21
Strodtbeck, F., 127

T
Tacit knowledge, 101, 104
Tanaka, K., 68
Tassoul, M., 47
Technological progress, 112
Telephone network, 139
The city of Kingship, 82
The Great, 78
The knowledge-creating company, 101
The Library of History, 79
The Life of Alexander, 79
The most famous office worker in the world, 68
The Parallel Lives, 79
The Persian Wars or the Greco-Persian Wars, 78
Thomas, J., 48
Three blades, 70
Torrance Tests of Creative Thinking (TTCT™), 36–37
Torrance, E.P., 36
Torrance, P., 36
Traditions, 13–15, 20, 90–92
Traits
  Are You a Crayon Breaker™, 37
  TTCT™ Traits, 36–37
Transcultural synergies, 2
Triandis, H., 127
Trompenaars, F., 152

U
Ulijin, 5
Uncertainty avoidance, 11, 14, 15, 18–20, 149
Uncertainty avoidance index (UAI), 149
Universalism, 2, 14, 20
Universalism–particularism, 127–129

V
Veblen T., 110, 111

W
Weiss, A., 8
Western culture, 53, 59
Western and non-Western civilizations, 13
Workplace, 89–90
Writing formats, 87, 88
Wyckoff, J., 12

X
Xu, F., 63–74

Y
Yaunā (Ionians), 78
Yuanzhang Zhu, 71